150
CONTRACTUAL PROBLEMS
AND THEIR SOLUTIONS

Roger Knowles
FRICS, FCIArb, FQSi, Barrister

Blackwell
Publishing

© 2005 by Blackwell Publishing Ltd

Editorial Offices:
Blackwell Publishing Ltd, 9600 Garsington Road, Oxford OX4 2DQ, UK
  Tel: +44 (0)1865 776868
Blackwell Publishing Inc., 350 Main Street, Malden, MA 02148-5020, USA
  Tel: +1 781 388 8250
Blackwell Publishing Asia Pty Ltd, 550 Swanston Street, Carlton, Victoria 3053, Australia
  Tel: +61 (0)3 8359 1011

The right of the Author to be identified as the Author of this Work has been asserted in accordance with the Copyright, Designs and Patents Act 1988.

First published as *One Hundred Contractual Problems and Their Solutions*, 2000 by Blackwell Science Ltd
Reprinted 2001 (three times), 2003

Library of Congress Cataloging-in-Publication Data
Knowles, Roger.
    150 contractual problems and their solutions/Roger Knowles.
      p. cm.
    Includes index.
    Rev. ed. of: One hundred contractual problems and their solutions. 2000.
    ISBN 1-4051-2070-3 (alk. paper)
    1. Contracts—Great Britain—Miscellanea.   I. Title: One hundred fifty problems and their solutions.
  II. Title.
  KD1554.Z9K58 2005
  346.4102—dc22

                                                                                    2004059597

ISBN 1-4051-2070-3

A catalogue record for this title is available from the British Library

Set in 10/12pt Palatino
by DP Photosetting, Aylesbury, Bucks
Printed and bound in
by Replika Press Pvt. Ltd, Kundli

The publisher's policy is to use permanent paper from mills that operate a sustainable forestry policy, and which has been manufactured from pulp processed using acid-free and elementary chlorine-free practices. Furthermore, the publisher ensures that the text paper and cover board used have met acceptable environmental accreditation standards.

For further information on Blackwell Publishing, visit our website:
www.blackwellpublishing.com

CONTENTS

Contents

PREFACE

A great deal has been happening in the field of construction law since the book was first published in 2000. So much so that the number of regular contractual problems has risen from 100 to 150, which is a surprising increase over a four year period.

The Housing Grants, Construction and Regeneration Act 1996 came into force on 1 May 1998. This legislation provides for adjudication to be available as of right to resolve disputes on all construction contracts. The adjudication process has become an instant success. There is a significant pool of adjudicators available who ensure that decisions are provided to the parties within the 28-day period following appointment as required by the Act which has greatly assisted in the success of the process. It has not, however, been achieved without difficulty. Challenges to the adjudicator's decisions by the losing party have been frequently referred to court.

The enforcement of an adjudicator's decision was the first matter to come before the courts. It had been suggested by many eminent lawyers that, if challenged, courts would not enforce an adjudicator's decision. Judges at the highest level soon decided that as it was the will of Parliament that adjudication should be the standard method of resolving disputes in the construction industry, then the decisions of adjudicators should be given legal backing.

The early case concerning whether decisions were to be enforced was quickly followed by many more legal challenges, one of the most notable being whether the courts would enforce an adjudicator's decision that was patently wrong. The Act only applies to construction contracts that are defined by the Act, but what is the position of those contracts where a part falls within the definition but the remainder does not? The purpose of the Act is to provide a process for resolving construction disputes, but how is a dispute defined? The contract to be caught by the Act must be in writing or evidenced in writing. What is meant by evidenced in writing? Adjudication is an interim measure whereby the adjudicator's decision is enforceable, but if one or both of the parties are dissatisfied the matter can be referred to arbitration or litigation at a later stage, which may result in the adjudicator's decision being reversed. If the losing party has to pay a sum of money in compliance with an adjudicator's decision, which due to the successful party's financial position may be almost impossible to recover in subsequent proceeding, can this be used as a legitimate reason for not complying with the adjudicator's decision? These and many more problems that have arisen as a result of the introduction of statutory adjudication are included in this edition of the book.

The Society of Construction Law published a Delay and Disruption Protocol in 2002. The object of the Protocol is to provide useful guidance on some of the common issues that arise on construction contracts. The guidance takes the form of statements of principle which affect many of the problems set out in this book. Some of these problems relate to:

- Procedure for granting extensions of time
- Float as it relates to time and compensation
- Concurrent delays
- Interest and finance charges
- Global claims
- Head office overhead formulae
- Disruption

The guidance provided by the Protocol has been added to those problems dealt with in the book which are relevant.

The construction industry has had to cope with many new ideas and procedures that have been introduced since the book was first published in 2000. Claims for negligence by contractors against architects whose contracts have been novated by an employer on a design and construct project have proved difficult to sustain. Clients who have insisted on contractors entering into contracts with a guaranteed maximum price can often be disappointed in efforts to resist requests for payment when the price rises above the guaranteed maximum. Retention has become something of a hot topic. How should contractors and subcontractors deal with abuses of the system? Contracts are now being let on an increasing scale whereby the contractor is reimbursed on a cost-recorded basis. A case arising from the tragedy of the foot and mouth epidemic has provided advice to contractors with regard to the retention of reliable records. Liability for consequential loss is often excluded by companies who regularly supply materials to the construction industry. What is meant by consequential loss? Have there been any recent developments to provide assistance in its interpretation? Courts have been under instruction to guide litigants in the direction of mediation. Does this create difficulties and how are they resolved? The legal system in the UK has in the past refused to enforce obligations for the parties to a contract to act in good faith, but is this changing? If an employer starts to occupy a building before all the work is fully completed does this constitute practical completion as far at the contract is concerned? These and many other problems resulting from changes and new ideas are dealt with in this edition.

Since 2000 new light has been shone on some of the old problems. The legal position of subsoil surveys, liability for payment in respect of speculative work undertaken in contemplation of receiving a contract, the effect of a lack of a delay notice on a contractor's entitlement to an extension of time, global claims, float in the programme, the effect of full and final settlement agreements, pay when paid clauses, employer's risk in agreeing to pay subcontractors direct, interpretation of reasonable costs, and liquidated damages have all been under review, resulting in either changes of reinforcement or existing positions. These are just some of the matters dealt with in this edition.

The changes outlined above convinced me of the need for this second edition.

Finally, my thanks again to Ann Glacki who, in like manner in the first edition, provided a great deal of the source material, and Suzanne Cash for word processing and reading over.

Roger Knowles

1
DESIGN

1.1 What is the difference between a fitness for purpose responsibility and an obligation to exercise reasonable skill and care?

1.1.1 Clients when appointing a designer, whether architect, engineer, contractor or subcontractor, expect the building or structure to operate when complete in the manner envisaged when the appointment was made. If the building or structure fails to meet the client's expectations there are often questions asked of the designer and/or contractor as to whether the problem resulted from a failure on their part to meet their contractual obligations. These obligations will normally take the form of implied or express terms in the conditions of appointment or the terms of the contract under which the work was carried out.

In the absence of an express term in the contract for providing a design service there will be an implied term that the designer will use reasonable skill and care. The standard is not that of the hypothetical 'reasonable man' of ordinary prudence and intelligence, but a higher standard related to his professed expertise. This was laid down in *Bolam* v. *Friern Hospital Management Committee* (1957) by Mr Justice McNair:

> [W]here you get a situation which involves the use of some special skill or competence, then the test whether there has been negligence or not is not the test of the man on top of a Clapham omnibus, because he has not got this special skill. The test is the standard of the ordinary skilled man exercising and professing to have that special skill A man need not possess the highest expert skill at the risk of being found negligent. It is well-established law that it is sufficient if he exercises the ordinary skill of an ordinary competent man exercising that particular art.

1.1.2 A person who professes to have a greater expertise than in fact he possesses will be judged on the basis of his pretended skills.

In *Wimpey Construction UK Ltd* v. *DV Poole* (1984), a case where unusually the plaintiffs were attempting to prove their own negligence, they attempted to convince the judge that a higher standard was appropriate to the case under consideration. They put forward two 'glosses', as the judge referred to them:

- First, that if the client deliberately obtains and pays for someone with specially high skill the Bolam test is not sufficient
- Second, that the professional person has a duty to exercise reasonable care in the light of his actual knowledge, not the lesser knowledge of the ordinary competent practitioner.

As regards the first gloss, the judge felt obliged to reject it in favour of the Bolam test. However the judge accepted the second gloss, not as a qualification of the Bolam test but as a direct application of the principle in *Donoghue* v. *Stevenson* (1932).

> You must take reasonable care to avoid acts or omissions which you can reasonably foresee would be likely to injure your neighbour.

1.1.3 Another important aspect of reasonable skill and care is what is generally referred to as the 'state of the art' defence. Briefly, what this means is that a designer is only expected to design in conformity with the accepted standards of the time. These standards will generally consist of Codes of Practice, British Standards or other authoritative published information.

1.1.4 Where a contractor or subcontractor undertakes design work or production of working drawings there is, in the absence of express term in the contract, an obligation to produce a product fit for its purpose. This is in marked contrast to a designer's implied obligation of reasonable skill and care.

The duty to produce a building fit for its purpose is an absolute duty independent of negligence. It is a duty which is greater than that imposed upon an architect employed solely to design, who would only be liable (in the absence of an express provision) if he were negligent. Express provisions to the contrary will obviously negate any implied terms.

The contractor's position is best illustrated by the following extracts from leading cases:

Independent Broadcasting Authority v. *EMI Electronics Limited* (1980)

> In the absence of a clear, contractual indication to the contrary, I see no reason why [a contractor] who in the course of his business contracts to design, supply and erect a television aerial mast is not under an obligation to ensure that it is reasonably fit for the purpose for which he knows it is intended to be used. The Court of Appeal held that this was the contractual obligation in this case and I agree with them. The critical question of fact is whether he for whom the mast was designed relied upon the skill of the supplier to design and supply a mast fit for the known purpose for which it was required ... In the absence of any terms (express or to be implied) negativing the obligation, one who contracts to design an article for any purpose made known to him undertakes that the design is reasonably fit for the purpose.

Greaves Contractors Limited v. *Baynham Meikle & Partners* (1975)

> Now as between the building owners and the contractors, it is plain that the owners made known to the contractors the purpose for which the building was required, so as to show that they relied on the contractors' skill and judgment. It was, therefore, the duty of the contractors to see that the finished work was reasonably fit for the purpose for which they knew it was required.

In the circumstances of this case, the designers were also held to have a liability to ensure that their design was fit for its purpose.

Young and Marten v. *McManus Childs* (1969)

> I think that the true view is that a person contracting to do work and supply materials warrants that the materials that he uses will be of good quality and reasonably fit for the purpose for which he is using them unless the circumstances of the contract are such as to exclude any such warranty.

1.1.5 The House of Lords' decision in *Slater* v. *Finning* (1996) held that no liability lies where a party is not aware of the particular purpose for which the goods are intended or where the proposed use deviates from the normal use.

The principle as expressed by Lord Keith was:

> 'As a matter of principle... it may be said that where a buyer purchases goods from a seller who deals in goods of that description there is no breach of the implied condition of fitness where the failure of the goods to meet the intended purpose arises from an abnormal feature or idiosyncrasy not made known to the seller by the buyer or in the circumstances of the use of the goods by the buyer. That is the case whether or not the buyer is himself aware of the abnormal feature or idiosyncrasy.'

Lord Steyn provided a useful example of the application of this decision in the construction industry in saying:

> 'If a contractor in England buys pipes from a dealer for use in a pipe-laying project the seller would normally assume that the pipes need merely to be suitable to withstand conditions in our moderate climate. If the contractor wishes to use the pipes in arctic conditions for a Siberian project, an implied condition that the pipes would be fit to withstand such extreme weather conditions could only be imputed to the seller if the buyer specifically made that purpose known to the seller.'

The decision was followed in *Q V Ltd* v. *Frederick Smith* (1998).

1.1.6 In the case of *PSC Freyssinet Ltd* v. *Bryne Brothers (Formwork) Ltd* (1996), the court had to decide liability where a design failure occurred due to a lack of provision for early thermal movement. The defendant, Byrne Brothers, was subcontractor for the design and construction of the car park superstructures at the Lakeside Shopping Complex in Thurrock, Essex. PSC was employed by Byrne Brothers to design and install post tensioned reinforcement and grouting. Whilst PSC owed a fitness for purpose obligation it was not responsible for the design of the whole beam as to do so would require a consideration of its relation to the entire structure and not merely the sub-frame. The court considered it absurd to hold PSC to a fitness for purpose term when their work might be affected by information supplied by a third party, namely the architect.

1.1.7 JCT With Contractor's Design in clause 2.5.1 places the following design responsibility upon the contractor:

> 'the Contractor shall have in respect of any defect or insufficiency in such design the like liability to the Employer, whether under statute or otherwise, as would an architect or ... other appropriate professional designer ...'

GC/Works/1 1998 under condition 10 (2) imposes a different responsibility. Condition 10 Alternative B states:

> 'The Contractor warrants to the Employer that any Works ... will be fit for their purposes, as made known to the Contractor by the Contract.'

ICE Design and Construct clause 8 (2) requires the contractor in carrying out his design responsibility to

> 'exercise all reasonable skill and care.'

It can be seen that some of the standard forms reduce the contractor's fitness for purpose obligation which the law would normally imply to the less onerous task of exercising reasonable skill and care. The main reason for this is the difficulty contractors have in obtaining insurance cover for a fitness for purpose obligation.

SUMMARY

In the absence of an express term in the conditions of contract a designer, whether architect, engineer or other designer, will have an implied obligation to carry out his design obligation employing reasonable skill and care. The test is whether the level of skill provided is the standard of the ordinary skilled person exercising and professing to have that skill.

Where a contractor or subcontractor undertakes a design responsibility in conjunction with an obligation to construct the works there is, in the absence of an express term in the contract, an implied obligation to produce a design which is reasonably fit for its purpose. This is an absolute duty and any failure of the design solution will place a responsibility upon the design and construct contractor or subcontractor whether or not the problem results from negligence.

Some of the standard forms reduce the contractor's fitness for purpose obligations which the law would normally imply to the less onerous reasonable skill and care.

1.2 Where a contractor/subcontractor's drawings are 'approved', 'checked', 'inspected', etc. by the architect/engineer and subsequently an error is discovered, who bears the cost – the contractor, subcontractor or employer? If the employer bears the cost can he recover the sum involved from the architect/engineer?

1.2.1   In general terms when an employer appoints an architect or engineer to design a building or work of a civil engineering nature, he is entitled to expect the architect or engineer to be responsible for all design work.

This basic principle was established in the case of *Moresk Cleaners Ltd v. Thomas Henwood Hicks* (1966).

The plaintiffs were launderers and dry cleaners who appointed the defendant architect to undertake the design work of an extension to their laundry. Instead of designing all the work himself, the architect arranged for the contractor to design the structure. The employer brought an action for defective design against the architect who argued that his terms of engagement entitled him to delegate the design of the structure to the contractor.

It was held that an architect has no power whatever to delegate his duty to anybody else. Sir Walter Carter QC had this to say:

> [Counsel for the architect] in a very powerful argument, asks me to say alternatively that the architect had implied authority to act as agent for the building owner to employ the contractor to design the structure and to find that he did just this. I am quite unable to accept that submission. In my opinion he had no implied authority to employ the contractor to design the building. If he wished to take that course, it was essential that he should obtain the permission of the building owner before that was done.

1.2.2 Nevertheless the architect or engineer in his terms of engagement may include a term which permits him to use a specialist contractor, sub-contractor or supplier to design any part of the works, leaving the architect or engineer with no responsibility if the design work undertaken by others contains a fault, but the employer has to agree to this.

Where a part of the design work is carried out by a subcontractor or supplier in accordance with an express term in the architect's or engineer's conditions of appointment it is in the employer's interests to obtain some form of design warranty from the subcontractor or supplier along the lines of NSC/W (for use where a nominated subcontractor undertakes design work under a JCT 98 contract). The employer would then be able to seek to recover any loss or damage resulting from design faults by the subcontractor or supplier on the basis of the warranty.

1.2.3 If, however, an architect or engineer (having excluded his responsibility for a subcontractor's design in the terms of his appointment) approves, checks or inspects a subcontractor's drawing, does he then take on any responsibility for any failure of the design?

It is essential for the architect or engineer to make it clear to both employer and subcontractor exactly what he is doing with the drawings. If he is checking the design carried out by the subcontractor or supplier he may find that, even though the terms of his appointment exclude liability, he may have adopted a post-contract amendment to the conditions and with it responsibility.

The employer will be left to bring an action against either the architect/engineer or the subcontractor who carried out the design. An unfortunate aspect of English law is that both may be held to be jointly and severally liable. In other words the employer can extract the full amount of his loss or damage from either party. This can be useful to the employer if a sub-contractor carried out the design and subsequently became insolvent, leaving a well insured architect who had checked the design to stand the full amount of the loss. Alternatively the employer may decide to sue both, leaving the court to allocate his loss or damage between the joint defendants after he has been paid in full by one or other of them.

1.2.4   If the architect/engineer is not checking the design then he must make it
very clear what he is doing. Ideally it should be set out in the architect's/
engineer's terms of appointment precisely what his duties are with regard to
design work undertaken by a contractor, subcontractor or supplier.

Should the employer commence an action against the architect/engineer
alone, then, under the Civil Liability (Contribution) Act 1978, the architect/
engineer may seek a contribution from the contractor, subcontractor or
supplier whose design was faulty. In the event of the employer deciding to
sue the contractor, subcontractor or supplier alone they, likewise, may seek a
contribution from the architect/engineer.

1.2.5   The fact that an engineer receives drawings does not in itself imply that he
has any liability for errors in design. In *J Sainsbury plc* v. *Broadway Malyan*
(1998) a claim for defective design was settled out of court. The problem
related to the design of a wall between a store area and retail area. Due to the
low level of fire protection fire spread and caused substantial damage. The
architect attempted to off-load some of the liability upon an engineer to
whom the drawings had been sent for comment. It was held that if the
architect wanted to get the structural engineer's advice on fire protection he
needed to say so. Simply to transmit the drawings for comment without
specifying any area in which comment was requested was not sufficient to
imply any obligation.

1.2.6   A different slant was placed upon acceptance of drawings by the engineer in
the case of *Shanks & McEwan (Contractors) Ltd* v. *Strathclyde Regional Council*
(1994) which arose out of the construction of a tunnel for a sewer. A method
of construction was employed using compressed air to minimise water
seepage. The tunnel and shaft segments in compliance with the specification
were designed by a supplier to the main contractor. The main contractor was
to be responsible for the adequacy of the design insofar as it was relevant to
his operations but it was also a requirement of the specification that design
calculations were to be submitted to the engineer. In the course of construction fine cracks appeared in the prefabricated tunnel segments due to a
design fault. The engineer was prepared to accept the work subject to the
segments being made reasonably watertight and confirmed the same in a
letter to the contractor dated 21 September 1990. Clause 8(2) of the ICE 5th
Edition which governed the contract states that the contractor shall not be
responsible for the design of the permanent works. There seemed to be a
conflict between clause 8(2) and the specification which placed responsibility for the design of the tunnel segments onto the contractor.

The contractor levied a claim for the cost of the repair work. It was the
view of the Court of Session in Scotland that, following acceptance by the
engineer of the design of the segments, the contractor was entitled to expect
that the approved design would not crack. The letter from the engineer
dated 21 September 1990 which accepted repair work to the segments was
held to be a variation and therefore the contractor was entitled to be paid for
that work.

1.2.7   The employer's ability to recover from the engineer any costs incurred
due to design error on the part of the contractor or subcontractor will
depend upon a number of factors. If the design faults lie with the

contractor or subcontractor it is to those who caused the error that the employer would normally address his claim. If the employer is unable to recover from the contractor or subcontractor, for example because of insolvency, he may wish to turn his attentions to the engineer. The ability to recover will depend upon the terms of the engineer's appointment. If the matter is referred to court all involved in the design process will normally be joined into the action.

In *London Underground* v. *Kenchington Ford* (1998) the design of a diaphragm wall at the Jubilee Line station of Canning Town became the subject of a dispute. The diaphragm wall was designed by Cementation Bachy (the contractors). London Underground argued that Kenchington Ford (the engineer) had failed to realise that there had been a mistake in computation made by Cementation Bachy and consequently the diaphragm wall was designed too deep and hence over expensive. The error had resulted from Cementation Bachy misinterpreting the load shown on the drawing. The contract stated that Cementation Bachy would be responsible for design errors whether approved by the engineer or not. Kenchington Ford were under a duty to London Underground to provide services which included the correction of any errors, ambiguities or omissions. The judge concluded that Kenchington Ford should have checked and discovered the error, and as they had not this constituted a breach of duty.

In *George Fischer (GB) Ltd* v. *Multi Design Consultants Roofdec Ltd, Severfield Reece and Davis Langdon and Everest* (1998), a complex multiparty action, the employer's representative was held to be partly liable in respect of the design error. The employer's representative's conditions of appointment obliged him to approve all working drawings. Following judgment in favour of the employer the parties agreed on the sum payable as damages. Multi Design Consultants, who carried out the design function, were liable in the sum of £940 000 with the liability of the employer's representative, Davis Langdon and Everest, being £807 388.

SUMMARY

The approval of a contractor or subcontractor's drawings by the architect or engineer will not usually relieve the contractor or subcontractor from liability. Employers who incur costs due to this type of error will normally commence an action against both the contractor/subcontractor who prepared the drawings and the architect/engineer who gave his approval. The court will decide on the apportionment of blame.

Where the employer incurs cost due to errors in the contractor/subcontractor's design, these costs may be recovered from the engineer/architect if a duty to check the drawing was expressly or impliedly provided for in the conditions of appointment and the errors result from a failure to carry out the checking properly.

In *Shanks and McEwan* v. *Strathclyde Regional Council* the design which proved faulty was approved by the engineer. The contractor was held to be entitled to recover from the employer the cost of remedial works. This case seems to cut across the accepted legal principle that one cannot benefit from one's own errors.

1.3 Who is responsible for co-ordinating design? Can a main contractor be legitimately given this responsibility even though he has no design responsibility?

1.3.1 As previously stated in 1.2.1, when an employer appoints an architect/engineer to design a building or work of a civil engineering nature he is entitled to expect the architect/engineer to be responsible for all design work.

This basic principle was established in *Moresk Cleaners Ltd* v. *Thomas Henwood Hicks* (1966).

1.3.2 This being the case, the architect/engineer will also be responsible for co-ordinating design unless there is an express term in the contract to the contrary.

1.3.3 Specifications for mechanical and electrical work and other specialist disciplines often refer to the subcontractor being responsible for design co-ordination. This will not absolve the architect from his design responsibilities expressed or implied in the conditions of engagement. If the specification which refers to a subcontractor being responsible for design co-ordination becomes a main contract document then the employer may bring an action against the main contractor for breach in respect of any loss or damage resulting from poor design co-ordination. Any liability on the part of the main contractor would be recoverable from the subcontractor under the terms of the subcontract. Alternatively, design co-ordination may be specifically referred to in a design warranty entered into by the subcontractor in which case the employer may commence an action upon the warranty against the subcontractor for faulty co-ordination. It is, however, for an architect/engineer to specify the obligation to co-ordinate in his conditions of engagement with the employer.

1.3.4 Where a JCT Design Portion Supplement is used, whereby the contractor is required to design a part of the work only, it will be the architect's responsibility to ensure that the contractor's design is properly co-ordinated with that of the architect.

1.3.5 The main contractor's responsibility for design co-ordination will be dependent upon the terms of the contract. Design by contractors, either employing a full design and construct procedure or a partial design and construct is on the increase. Even without a design responsibility the terms of the main contract may impose a responsibility upon the main contractor to undertake design co-ordination. However, it is unlikely that, in the absence of express terms in a main contract or subcontract, an obligation to co-ordinate design will rest on the main contractor or subcontractor.

1.3.6 If the contractor is required to co-ordinate design work an express clause must be included in the contract which is fully descriptive of the co-ordinating activities required of the contractor. A brief term which states that the contractor is responsible for co-ordinating the work of all subcontractors including design would not be adequate. A much more descriptive clause is necessary. This clause should indicate which trades are involved and expressly state that all costs and losses resulting from a failure properly to co-ordinate the subcontractor's design and working drawings will be borne by the main contractor.

1.3.7   Where there is no reference to a design obligation in the main contract, it is unlikely that the main contractor will become liable for any defective design by a subcontractor: *Norta* v. *John Sisk* (1971).

SUMMARY

The architect/engineer will normally be responsible for design co-ordination except where the contractor is appointed on a design and construct basis. It is possible for an architect to disclaim the responsibility for design co-ordination in his conditions of engagement with the employer and place the burden upon the contractor's shoulders. For a main contractor to take on a responsibility for design co-ordination will require a fully descriptive clause.

1.4   Can a contractor be held responsible for a design error where the employer appoints an architect and no provision exists in the contract for the contractor to undertake any design responsibility?

1.4.1   It is commonplace for a contractor to have placed upon him by the terms of contract a full design responsibility. Some contracts provide for parts only of the work to be designed by the contractor. If under the contract the employer appoints an architect whose duty it is to prepare all the drawings with no reference being made to a contractor's design responsibility, can a situation ever arise where the contractor finds himself liable for a design fault?

1.4.2   In the case of *Edward Lindenberg* v. *Joe Canning, Jerome Contracting Ltd* (1992) the plaintiff engaged the defendant builder for some conversion work on a block of flats. During the work load-bearing walls in the cellar were demolished which caused damage in the flat above. The plaintiff sued the defendants for breach of contract and/or negligence, seeking repayment of the sums he was forced to pay the building owners under an indemnity.

The plaintiff alleged that Canning was in breach of an implied term that he would proceed in a good and workmanlike manner, and that he had negligently demolished the load-bearing walls without providing temporary or permanent support.

It was held:

(1)   As there was no express agreement between the parties, Canning was entitled to be paid on a quantum meruit basis for labour and materials.

(2)   There was an implied term that the defendant would undertake the

work in a good and workmanlike manner and exercise the care expected of a competent builder. He had been supplied with plans, prepared by the plaintiff's surveyor, which supposedly indicated which walls were non load-bearing. However, as a builder, he should have known that since they were nine inch walls they were in fact load-bearing. As he took 'much less care than was to be expected of an ordinary competent builder' he was in breach of contract but not liable in negligence.

(3)   The plaintiff was entitled to recover £7484 (representing the amount he had to reimburse the building owner plus professional fees) less a sum for contributory negligence.

(4)   The plaintiff had been guilty of contributory negligence through his agents in that Canning had been given plans which wrongly showed which walls were non load-bearing, oral instruction had been given to demolish walls and no instructions had been given regarding the provision of supports. The liability was attributed at 75% to the plaintiff and 25% to the defendant. The plaintiff's damages were reduced accordingly to £1871.

(5)   Canning was entitled to a quantum meruit payment, assessed at £4893. As this was less than the £7000 which the plaintiff had advanced to him, Canning was liable to repay the difference.

SUMMARY

The fact that the employer employs an architect and the main contract makes no reference to the contractor's design responsibility does not mean that the contractor cannot become responsible for design errors. In the *Joe Canning* case the drawings incorrectly showed which walls were load-bearing. The contractor was nevertheless held to be liable in breach of contract for taking much less care than an ordinary competent builder in demolishing the walls which turned out to be load-bearing.

1.5   Can a main contractor be responsible if a nominated/named subcontractor's design is defective?

1.5.1   Whether a main contractor is responsible for a nominated or named subcontractor's design error is usually decided following a careful study of the contract documents. It is common practice for the architect or engineer to arrange for specialist work to be designed by a subcontractor who is then either nominated or named in the contract documents. Often the main contractor has no involvement whatsoever in the design of the specialist work.

1.5.2   The matter is catered for in the ICE 6th and 7th Editions at clause 58(3) which states:

'If in connection with any Provisional Sum or Prime Cost Item the services to be provided include any matter of design or specification of any part of the Permanent Works or of any equipment or plant to be incorporated therein such requirement shall be expressly stated in the Contract and shall be included in any Nominated Sub-contract. The obligation of the Contractor in respect thereof shall be only that which has been expressly stated in accordance with this sub-clause.'

The ICE contracts therefore make it crystal clear as to where the contractor's responsibility lies with regard to the design of a nominated subcontractor's work.

1.5.3　JCT 98 in clause 35.21, in like manner to the ICE 6th and 7th Editions, makes it clear that the contractor is not responsible for design work undertaken by a nominated subcontractor.

1.5.4　In the case of *Norta* v. *John Sisk* (1977) the Irish Supreme Court had to decide the contractor's liability for a nominated subcontractor's design error where the conditions of the main contract made no reference to design responsibility. The claimant entered into a contract to construct a factory for making wallpaper. Prior to the receipt of tenders from main contractors the claimant approved a quotation from Hoesch Export for the design, supply and erection of the superstructure of the factory including roof lights. Hoesch Export became nominated subcontractors to John Sisk, the appointed main contractor. Following practical completion the roof began to leak due to faulty design of the roof lights. The claimant sought to recover his losses from the main contractor John Sisk. No reference was made in the main contract to John Sisk having any design responsibility. It was argued on behalf of the claimant that a design obligation was implied into the main contract. The Irish Supreme Court held that no such term could be implied into the main contract and therefore John Sisk had no liability.

1.5.5　JCT 98 includes for performance specified work. Clause 42 provides for performance specified work to be included in the contract by means of the employer indicating the performance he requires from such work. Before carrying out the work the contractor must produce a contractor's statement in sufficient form and detail adequately to explain the contractor's proposals.

　　The contractor will be responsible for any fault in the contractor's statement, which may include design work by subcontractors if the fault results from a failure to exercise reasonable skill and care.

1.5.6　The main contractor will be responsible for all design work including that of subcontractors where design and construct conditions apply, e.g. JCT With Contractor's Design.

1.5.7　Many non standard forms of contract or amendments to standard forms make it clear that the main contractor is responsible to the employer for all the nominated subcontractor's work including design.

SUMMARY

The main contractor will be responsible for design faults in a named or nominated subcontractor's work if there is a clear statement to that effect in the main contract. In the absence of an express obligation an employer would have to show that such an obligation was implied. This may prove difficult if the subcontractor's design work was developed through a liaison between the subcontractor and architect/engineer direct, particularly if this took place without any involvement by the contractor. To protect himself against loss due to subcontractor's design faults it is advisable for the employer to enter into a design warranty direct with the subcontractor. Most of the commonly used standard forms of contract make it clear that the main contractor is not responsible for a nominated subcontractor or nominated supplier's defective design.

1.6   Must a contractor notify an architect/engineer of defects in his design?

1.6.1   Human errors occur on a regular basis including design errors by architects and engineers. Contractors may suspect that a design error has occurred. If this be the case does the contractor have an obligation to draw attention to the design error?

1.6.2   In the case of *Equitable Debenture Assets Corporation Ltd* v. *William Moss and Others* (1984) it was held that a term should be implied into the contract that the contractor should report design defects known to him.

1.6.3   The case of *Victoria University of Manchester* v. *Hugh Wilson and Others* (1984) dealt with a problem of ceramic tiles falling off the exterior face of a building at Manchester University (see 1.7.2). The cause was a combination of poor design and poor workmanship. With regard to design defects it was held that the contractor had a duty under an implied term of JCT 63 to warn of design defects which they believed to exist but with no obligation to make a close scrutiny of the architect's drawings. Judge John Newey said:

> The contractor's duty to warn the architect of defects which they believe existed in the architect's design did not in my view require them to make a critical survey of the drawings, bills and specifications looking meticulously for mistakes.'

1.6.4   A more recent decision is *University of Glasgow* v. *Whitfield and Laing* (1988) which called into question the decisions in *Equitable Debenture Assets Corporation* and *Victoria University*. In this case it was alleged that the contractor owed an implied duty to the architect to warn of design faults.

However Judge Bowsher had this to say when holding that the contractor had no duty to the architect to warn of defects:

> Mr Gaitskell on behalf of the defendant relies on the decisions of Judge Newey QC in *Equitable Debenture Assets Corporation* v. *William Moss* (1984) and *Victoria University of Manchester* v. *Wilson* (1984). On analysis it is clear that both cases were concerned with a duty of a contractor to warn the employer, not a duty owed to the architect to warn the architect. References to a duty to give a warning to the architect were in both cases references to a duty to warn the architect as agent of the employer. It is clear from page 163 of the report of the *Victoria University of Manchester* case that the learned judge considered that both decisions were founded on implied contract between the contractor and the building owner. In each case, the learned judge cited *Duncan* v. *Blundell* (1820) and *Brunswick Construction Limited* v. *Nowlan* (1974). It is plain from the citation from the *Brunswick Construction* case that the learned judge had in mind the situation where the contractor knew that the owner placed reliance on him in the matter

of design. It seems to me that the decisions in *EDAC* v. *Moss* and *Victoria University of Manchester* can stand with more recent decisions if they are read as cases where there was a special relationship between the parties, but not otherwise, and bearing in mind the difficulties in analysing the meaning of the words 'special relationship' and 'reliance' demonstrated by Robert Goff LJ in *Muirhead* v. *Industrial Tank Limited* (1986). On the facts of the present case it is not necessary to resolve those difficulties.'

1.6.5 In *Edward Lindenberg* v. *Joe Canning Jerome Contracting Ltd* (1992) (see 1.4.2) the contractor was held liable to make a contribution to the cost of remedial works resulting from the demolition of load-bearing walls. The walls were shown on the architect's drawings as non load-bearing.

1.6.6 The subject of a contractor's duty to warn occurred in *Plant Construction plc* v. *Clive Adams Associates and Another* (1999), heard before the Court of Appeal. Ford appointed a company trading under the name of Plant to install two engine mounts in a research and development centre at Laindon, Essex.

The substructure and underpinning of a roof was subcontracted to JMH. A variation was issued involving the design of the temporary works by a representative of Ford. The design was defective and collapse occurred. JMH was not responsible for the design but it was held that they had an implied obligation to exercise reasonable skill and care. An experienced contractor such as JMH would have an obligation to warn of errors in design which were obviously dangerous and defective. The decision left open the situation where the design is obviously defective but not dangerous.

1.6.7 Some forms of contract, for example JCT 98 clause 2.3, require the contractor to notify the architect of any discrepancy in any drawings issued by the architect.

SUMMARY

It was held in the *University of Glasgow* v. *Whitfield and Laing* case that in the absence of express provisions the contractor may have an implied duty to the employer to warn of design faults only where a special relationship exists between them. There would otherwise appear to be no obligation in the absence of an express term in the contract.

This decision is difficult to comprehend. If correct a contractor, knowing of a design error, could carry out construction work without obligation. It is hard to anticipate any subsequent cases following this decision. The decision in the *Equitable Debenture* case is to be preferred, where it was held that an implied term exists in construction contracts that contractors should report design defects known to them. In the case of *Plant Construction plc* v. *Clive Adams* (1999) it was held that a contractor would have an obligation to warn of errors in design which were obviously dangerous and defective.

1.7   Where an architect/engineer includes a new product in his design following advice from a manufacturer and the product proves to be unsuitable, is the architect/engineer liable to the employer for his losses?

1.7.1   Engineers and architects often have difficulty in providing appropriate design solutions to suit planning constraints, environmental considerations and the client's financial position. Manufacturers often make claims that a new product will meet the architect/engineer's requirements. In the absence of a track record the architect/engineer is seen to be taking a risk in specifying the new product. If, having made as many checks concerning the manufacturing process and having sought whatever advice is available, the architect/engineer specifies the product, what liability does the architect have to the client if the product proves unsatisfactory?

1.7.2   The case of *Victoria University of Manchester* v. *Hugh Wilson and Others* (1984) arose out of a major development for the plaintiffs erected in two phases between 1968 and 1976. The first defendants were the architects for the development, the second defendants the main contractors and the third defendants nominated subcontractors. The architects' design called for a building of reinforced concrete (which was not waterproof) to be clad partly in red Accrington bricks and partly in ceramic tiles. In due course many of the tiles fell off and the University adopted a remedial plan which involved the erection of brick cladding with a cavity between bricks and tiles and with the brick walls attached to the structure by steel ties.

It was held that the architect was liable as his design was defective. With regard to the use of untried materials Judge John Newey had this to say:

> For architects to use untried, or relatively untried materials or techniques cannot in itself be wrong, as otherwise the construction industry can never make any progress. I think, however, that architects who are venturing into the untried or little tried would be wise to warn their clients specifically of what they are doing and to obtain their express approval.

1.7.3   In *Richard Roberts Holdings Ltd* v. *Douglas Smith Stimson Partnership* (1988) a tank lining failed. The employer brought an action against the architect for negligence. The architect's defence was that he had no legal liability as the employer knew that he had no knowledge of linings. It was held, again by Judge John Newey, that:

> The architects were employed for the design of the whole scheme of which the linings were an integral part. The architects did not know about linings, but part of their expertise as architect was to be able to collect information about materials of which they lacked knowledge and/or experience and to form a view about them. If the architects felt that they could not form a reliable judgment about a lining for a tank they should have informed the employer of that fact and advised them to take other advice . . .

SUMMARY

Where an engineer/architect includes a new product in his design the employer should be informed at the outset. Failure to advise the employer would leave the engineer/architect exposed to a liability for negligence should the new product fail.

1.8 Where an architect/engineer approves or accepts a subcontractor's drawings, how long can he take before an entitlement to an extension of time arises?

1.8.1 It is quite common for subcontractors to be required to produce drawings in respect of their installation. Well drafted specifications will normally provide for an approval or acceptance system. The system will set out the roles to be played by architect/engineer and subcontractor up to the stage of approval or acceptance of the drawings. Usually a time scale will be included which will indicate the maximum time within which the drawings must be approved or accepted or queries raised. Time will normally be allowed for answering queries with final approval or acceptance, again within a timescale. If the architect or engineer fails to approve, accept or query a subcontractor's drawing within the timescale, and as a result the completion date for the project is delayed, there is usually an entitlement to an extension of time. If there is no provision for extending time where delays are caused by late approval or acceptance then time becomes at large and the subcontractor's obligation is to complete within a reasonable time.

1.8.2 GC/Works/1 Design and Build 1998 requires the contractor to ensure that the programme allows reasonable periods of time for the provision of information from the employer.

1.8.3 Contractors and subcontractors will often indicate on the face of the drawing a period of time within which approval is sought.

1.8.4 Where there is no timescale in the procedures within which the architect/engineer is required to approve or accept or query a subcontractor's drawing, or perhaps there is no formal procedure provided for approvals in the specification, the court will normally hold that such a term will be implied to give the contract business efficacy. A clause will usually be implied to the effect that approval by the architect/engineer must be given or any query raised within a reasonable time. What is a reasonable time will depend upon the circumstances of each case and would include such matters as any time allowed on the subcontractor's programme, the rate of progress of the work and the date fixed for completion.

SUMMARY

Ideally the contract will indicate how long is provided for drawing approval or the contractor's programme should address the point. If there is no provision in the contract then it will be implied that a reasonable period will be allowed.

1.9 Where is the line to be drawn between an architect/engineer's duty to design the works or a system and a contractor or subcontractor's obligation to produce working shop or installation drawings?

1.9.1   Where a contract such as JCT 98, ICE 6th or 7th Edition, MF/1 or the GC/Works/1 1998 is employed, the duty to design the works rests with the architect/engineer. However provision is made in these contracts for some of the design work to be prepared by the contractor. Many bespoke engineering contracts require the contractor to be responsible for the detailed design of the plant and of the works in accordance with the specification. Specifications are often written to the effect that specialist engineering subcontractors will be obliged to produce shop or working drawings. There is no hard and fast rule as to where the engineer's obligations cease and those of the contractor or engineering subcontractor begin. It will be a matter for decision in each and every case.

1.9.2   In *H. Fairweather & Co* v. *London Borough of Wandsworth* (1987), a subcontract was let using the now out of date NFBTE/FASS nominated subcontract often referred to as the Green Form. The description of the works set out in the appendix to that form was to 'carry out the installation and testing of the underground heat distribution system as described in [the specification]'.

The specification had two provisions. Clause 1.15 made it the subcontractor's responsibility to provide the installation drawings and they were also 'responsible for providing all installation drawings in good time to meet the agreed programme for the works'. Section 3(b) of the technical specification also required detailed drawings to be prepared and supplied by the subcontractor.

Before entering into the nominated subcontract Fairweathers had written to the architect in an endeavour to disclaim 'any responsibility for the design work that may be undertaken by your nominated subcontractor'. They also asked for 'a suitable indemnity against defects in design work carried out by the nominated subcontractor'. The architect's reply drew attention to the provisions of clause 1.15 and pointed out that these did not 'require [them] to assume responsibility for the design of the system' Fairweathers did not take the matter further and entered into the subcontract.

The arbitrator found that the installation drawings were not design drawings. The judge agreed with him although he had not seen the drawings. It does not appear that there was any dispute about responsibility for the content of the installation drawings and it would seem from this case that one cannot deduce that 'installation drawings' in general do not embody any 'design'. The architect had made it clear that the installation drawings were to be provided so as to meet the requirements of the programme and that the subcontractors were not responsible for the design of the system. However, in the course of preparing a detailed design for the installation of a system decisions are taken of a design nature by the person responsible for the preparation of the drawings. In the absence of a clear contrary indication the responsible contractor, subcontractor or supplier will be held liable.

1.9.3 It is not always obvious where the line is to be drawn between design or conceptual design and shop or working drawings. What is the purpose of the shop or working drawings? Some may argue that the intention is that the contractor or subcontractor's duty is to fill in the gaps left in the design or conceptual design drawings. Others may argue that the purpose of shop or working drawings is to convert design information into a format to enable the materials to be manufactured and fixed.

1.9.4 It is essential if a specified or nominated subcontractor is to produce shop or working drawings for the contract to stipulate in clear terms what is meant by these terms.

SUMMARY

It would seem that it is impossible to produce a dividing line to differentiate between design drawings and working shop or installation drawings. Each case would have to be judged on its merits. A reasonable interpretation is that the purpose of shop or working drawings is to convert design information into a format to enable the materials to be manufactured and fixed.

1.10 Where an item of work has been properly provided for in the Employer's Requirements but is missing from the Contractor's Proposals, can the contractor claim extra payment for doing the work on the grounds that it was never included in the contract price?

1.10.1 If we were living in a perfect world then all contract documents would be fault free. Unfortunately human beings are often known to err and as a result discrepancies are apt to appear between the employer's requirements and contractor's proposals.

1.10.2 The recitals to the JCT With Contractor's Design state:

'the Employer is desirous of obtaining the construction of the following works ... for which works he has issued to the contractor his requirements (hereinafter referred to as 'the Employer's Requirements'); ...'

'The Contractor has submitted proposals for carrying out the works ... (hereinafter referred to as 'the Contractor's Proposals') ...'

The contractor's obligations are set out in clause 2.1 and expressed in the following terms:

'The Contractor shall upon and subject to the Conditions carry out and complete the Works referred to in the Employer's Requirements, the Contractor's Proposals (to which the Contract Sum Analysis is annexed), the Articles of Agreement, these Conditions and the Appendices in accordance with the aforementioned documents.'

1.10.3 In anticipation of discrepancies and conflicts arising it is usual for contracts to lay down rules as to how they should be adjusted. JCT With Contractor's Design deals with discrepancies in the following manner:

- Clause 2.2 stipulates that nothing contained in the employer's requirements or the contractor's proposals or the contract sum analysis can override or modify the application or interpretation of the printed form. It is, therefore, important that any amendment to be made to the contract is made to the printed form because it is clear that a term in either the employer's requirements or the contractor's proposals which attempts to amend the printed form will be ineffective.
- Under clause 2.4.1 if there is any discrepancy in the employer's requirements, the contractor's proposals (if they deal with the discrepancy) will prevail without adjustment to the contract sum. If, however, they do not deal with the discrepancy, the contractor must submit written proposals. The employer then has a choice. He can agree the proposals or himself decide on another course of action. Either case ranks as a change in the employer's requirements for payment if appropriate.
- Clause 2.4.2 comes into play if there is any discrepancy in the contractor's proposals. The contractor must submit his proposals to overcome the discrepancy and the employer may choose between the proposals and the discrepant items. The contractor must comply without extra cost to the employer.

1.10.4   A more difficult situation arises if there is a discrepancy between employer's requirements and contractor's proposals. This is a common occurrence in practice, for example the employer's requirements may call for engineering bricks below the damp proof course whereas the contractor's proposals allow for semi-engineering bricks, either type of brick being suitable.

The contract is silent as to how this type of discrepancy is to be dealt with, but a clue as to how the situation can be resolved is contained in the third recital which states:

> 'The Employer has examined the Contractor's Proposals and the Contract Sum Analysis and, subject to the Conditions and, where applicable the Supplementary Provisions hereinafter contained, is satisfied that they appear to meet the Employer's Requirements.'

It is arguable that, as the employer has declared that he has examined and is satisfied with the contractor's proposals, any discrepancy between the employer's requirements and contractor's proposals which comes to light after the contract has been entered into should be interpreted in the contractor's favour. There is however no authority for this argument.

An amendment to the wording of the third recital should be made to indicate which takes precedence.

1.10.5   The GC/Works/1 1998 Design and Build contract is reasonably clear as to which of the employer's requirements or the contractor's proposals takes precedence.

Condition 2 (2) states:

> 'In the case of discrepancy between the Employer's Requirements and either the Contractor's Proposals or the Pricing Document, the Employer's Requirements will prevail without adjustment to the Contract Sum'.

Further references to discrepancies are made in condition 10A which states:

> 'To demonstrate compliance with the Employer's Requirements the contractor shall ensure that relevant work will be the subject of a Design Document.'

Condition 10A(7) develops the theme further by stating that:

> 'In case of any discrepancy between Employer's Requirements and Design Documents the Employer's Requirements shall prevail, without any adjustment to the Contract Sum.'

'Design documents' are defined as any drawing, plan, sketch, calculation, specification or any other document prepared in connection with design by the contractor. The intention is to catch any document whether prepared prior to the submission of the tender or subsequently prepared by the contractor for design purposes. All these documents will be subsequent to the employer's requirements.

1.10.6 The ICE Design and Construct Conditions are also clear as to the priority of those key documents in that clause 5(b) states:

> 'If in the light of the several documents forming the Contract there remain ambiguities or discrepancies between the Employer's Requirements and the Contractor's Submission the Employer's Requirements shall prevail.'

1.10.7 A court may take the view that whilst the Employer's Requirements and Contractor's Proposals are silent with regard to a particular item of work, its need was obvious (*Williams* v. *Fitzmaurice* (1858)). For example, a door will always require ironmongery and a house will require flooring.

SUMMARY

Unfortunately the JCT With Contractor's Design contract does not address the difficulty which may arise where there is a conflict between the employer's requirements and the contractor's proposals. The contract is silent as to which will take precedence. It is likely, however, that a court would hold that the contractor's proposals take precedence as the third recital indicates that:

> 'the Employer has examined the Contractor's Proposals ... and is satisfied that they appear to meet the Employer's Requirements.'

The ICE Design and Construct and GC Works/1 Design and Build contracts make it clear that the employer's requirements will take precedence over the contractor's proposals.

1.11 Is the contractor entitled to payment for design in full when the design work has been completed or should payment for design costs be spread over the value of work as and when it is carried out?

1.11.1 Contracts which are well drafted will usually be precise as to how much is to be paid or the manner in which payment is to be calculated and the timing of the payment. Design and construct contracts are no exception, and so the contract should be clear as to when payment for both the design function and construction of the works is to be made.

1.11.2   Contracts such as GC/Works/1 Design and Build provide for milestone payments. This being the case the milestone payment chart should make it clear when payment for design is to be made. In considering the make up of each payment consideration should be given to the contractor's pre-contract and post-contract design costs. Provision for payment of the pre-contract design costs should be included in the first milestone. The post-contract costs should be costed in accordance with a design programme and allocated to the appropriate milestone. Condition 48B provides for mobilisation payment if stated in the abstract of particulars. The calculation of this payment would normally include pre-contract design costs. Where milestone payment and mobilisation payment provisions do not apply payment of the pre-contract design costs should be included in the first advance on account. Design costs should be included in subsequent advances on account to accord with the progress of the post contract design.

1.11.3   JCT With Contractor's Design is similar to GC/Works/1 1998. Clause 30.2 A, Alternative A, provides for stage payments. The analysis of stage payments included in the appendix should make it clear in which stage the pre-contract and post-contract design costs will be paid. If Alternative A does not apply Alternative B comes into operation. In this case payment of the pre-contract design costs should be included in the first interim payment and the remaining design costs to be included in subsequent payments to suit the progress of the design.

In like manner to GC/Works/1 1998, JCT With Contractor's Design provides an option for an advanced payment to be made. Such advanced payment would normally include pre-contract design costs.

1.11.4   ICE Design and Construct makes provision in clause 60(2)(a) for a payment schedule to be included in the contract. This schedule should make it clear as to when payment for pre-contract and post-contract design costs are to be made. If there is no schedule design costs should be dealt with in the same manner as Alternative B of JCT With Contractor's Design, 1998 edition. No provision is made for advance payment or mobilisation payment.

SUMMARY

Payments should reflect the fact that design costs comprise pre-contract design costs and post-contract design costs. Where stage or milestone payments apply these costs should be properly allocated to the appropriate stage or milestone. The first one should include for pre-contract design costs. If there is no provision for stage or milestone payments the first interim payment should include pre-contract design costs. The post-contract design costs should be included in subsequent interim payments to suit the progress of the design.

1.12 On a design and construct project where the architect is novated from the employer to the contractor, is there any impediment upon the contractor's ability to recover from the architect loss he suffers due to architect design errors which occurred during his employment by the employer?

1.12.1 It has become a common practice for employers wishing to enter into a design and construct contract to start off the process by appointing an architect themselves. The intention is for the architect to be involved in the planning application and to work up the design to a state where tenders from contractors can be sought. When the contractor is appointed, the architect, by way of a novation agreement, becomes a part of the contractor's team. Under the novation agreement the contractor takes responsibility for the work carried out by the architect both pre and post contract.

1.12.2 The wording of the novation agreement provides for the contractor to stand in the employer's shoes with regard to negligence on the part of the architect. Any right of redress vested in the employer regarding the negligence of the architect in the pre-contract stage is transferred to the contractor. Contractors have derived comfort from this arrangement. They considered that any loss incurred as a result of an architect's pre-contract error could be recovered from the architect.

1.12.3 In the case of *Blyth and Blyth* v. *Carillion Construction Ltd* (2001), the architect was responsible for design errors in the pre-contract stage which resulted in the contractor incurring additional cost. The novation agreement allowed the contractor to pursue claims against the architect which would have been available to the employer. In other words, whatever loss the employer would have incurred resulting from the design errors was recoverable by the contractor. The design errors affected the contractor's price but would not have involved the employer in any additional cost. The contractor therefore recovered nothing. This will come as a shock to regular design and construct contractors. Serious rewriting of novation agreements was obviously necessary, as a result of this decision.

SUMMARY

Under a design and construct contract, the contractor takes responsibility for all design work carried out both pre and post contract. The intention of a novation agreement is to transfer the architect design obligation in the pre-contract stage from the employer to the contractor.

Comfort can be drawn by the contractor from the novation agreement concerning design errors due to negligence by the architect which occurred in the pre-contract stage. Whilst the contractor can be held responsible to the employer for the design errors, redress by the contractor can be sought from the architect. Unfortunately, due to the particular wording of many novation agreements, as the employer is unlikely to suffer loss due to an architect's negligence, as the risk has been transferred to the contractor, he in turn can recover nothing from the architect.

The CIC/Nov Agr novation agreement published in 2004 by the Construction Industry Council deals with this problem. In clause 4(a) the consultant warrants to the contractor that all services provided to the employer have been performed in accordance with the terms of the original appointment.

2
TENDERS

2.1 Where an employer includes with the tender enquiry documents a site survey which proves misleading, can this be the basis of a claim?

2.1.1 Problems often arise where unforeseen adverse ground conditions occur which add to the contractor or subcontractor's costs. Who pays the bill?

2.1.2 The ICE 6th Edition clause 11(3) deals with the matter in precise terms where it states:

'The contractor shall be deemed to have

(a) based his tender on the information made available by the Employer and on his own inspection and examination all as aforementioned.'

It would seem that where the ICE conditions apply the employer is taking the responsibility for the accuracy of the information he provides at tender stage.

The ICE 7th Edition is worded in a slightly different manner where it states in clause 11(3)

'The Contractor shall be deemed to have ... based his tender on his own inspection and examination as aforesaid and on all information whether obtainable by him or made available by the Employer ...'

The intention of the revised wording is that the contractor's tender is deemed to be based upon not only information provided by the employer or as a result of his own inspection but also information derived from other sources. This could include information obtained from utilities such as water or gas companies.

2.1.3 The Engineering and Construction Contract (or New Engineering Contract, as it is still often called) provides in clause 60.1 for the contractor to recover time and cost where physical conditions are encountered which the contractor would have judged at the contract date to have had a small chance of occurring. In judging physical conditions the contractor, in accordance with clause 60.2, is assumed to have taken into account the site information which will normally be provided by the employer.

2.1.4 GC/Works/1 1998 condition 7(1) requires the contractor to have satisfied himself, among other matters, as to the nature of the soil and materials to be excavated. Condition 7(3) goes on to say that if ground conditions are encountered which the contractor did not know of and which he could not reasonably have foreseen (having regard to any information which he had or ought reasonably to have ascertained) he will become entitled to claim extra.

2.1.5   Under the Misrepresentation Act 1967 a subsoil survey which proves to be misleading though innocently made could give rise to a claim for damages. A good defence for the employer is however, available if he can demonstrate that he had reasonable grounds to believe and did believe that the information contained in the subsoil survey was correct.

A disclaimer will be of no effect unless it can be shown to be fair and reasonable having regard to the circumstances which were or ought reasonably to have been known or in the contemplation of the parties when the contract was made.

Max Abrahamson in his book *Engineering Law and the ICE Contracts* says:

> 'The courts are obviously disinclined to allow a party to make a groundless misrepresentation without accepting liability for the consequences.'

For example, in *Howard Marine & Dredging Co Ltd* v. *A Ogden & Sons (Excavations) Ltd* (1978) owners of a barge were held liable to contractors who had hired the barge for construction works on the faith of a misrepresentation of the barge's dead weight even though the charterparty stated that the charterers' 'acceptance of handing over the vessel shall be conclusive that they have examined the vessel and found her to be in all respects ... fit for the intended and contemplated use by the charterers and in every other way satisfactory to them. The defendant's marine manager had said at a meeting that the payload of the barge was 1600 tonnes, whereas in fact it was only 1055 tonnes. The mis-statement was based on the manager's recollection of a figure given in Lloyds Register that was incorrect. He had at some time seen shipping documents which gave a more correct figure but that had not registered in his mind.

2.1.6   In *Pearson and Son Ltd* v. *Dublin Corporation* (1907) the engineer had shown a wall on the contract drawings in a position which he knew was not correct. There was a clause in the contract to the effect that the contractor would satisfy himself as to the dimensions, levels and nature of all existing works and that the employer did not hold himself responsible for the accuracy of information given. Nevertheless this was held to be no defence to an action for fraud.

Alternatively the contractor who is misled may have a remedy in tort for negligent mis-statements on the principle of *Hedley Byrne & Co Ltd* v. *Heller & Partners* (1963).

2.1.7   The question of misrepresentation under the Misrepresentation Act 1967 and negligent mis-statement was the subject of the decision in *Turriff Ltd* v. *Welsh National Water Authority* (1979). An issue in the case related to an alleged misrepresentation in the specification, in which it was stated that satisfactory test units had already been carried out by Trocoll Industries. It was also stated that if competently incorporated into the works the units could achieve a standard of accuracy within the desired tolerances and also that all the features of the units were proven. These representations were found by the court to be incorrect.

The judge found that Turriff had relied upon the statements and had entered into the contract on the understanding that the representations were true. It was found by the court that the employer was unable to prove there were reasonable grounds for believing the representations to be true and therefore, in accordance with the Misrepresentation Act 1967, Turriff was entitled to the damages they had suffered as a result.

The court also held that in so far as the representations contained in the specification were statements of opinion and not fact (and thus not within the gambit of misrepresentation) they in any event amounted to negligent mis-statements. In line with the decision in *Hedley Byrne & Co Ltd* v. *Heller & Partners Ltd* (1963) Turriff was entitled to damages.

While the representations in the specification concerning the units are peculiar to this particular contract, a question of misrepresentation and negligent mis-statement may often be relevant in relation to engineers providing subsoil survey information to tendering contractors which proves to be inaccurate.

2.1.8 Contractors who suffer from incorrect subsoil surveys may be able to demonstrate that the survey became a condition or warranty in the contract as was the situation in the case of *Bacal Construction* v. *Northampton Development Corporation* (1975). The contractor had submitted as part of his tender sub-structure designs and detailed priced bills of quantities for six selected blocks in selected foundation conditions. The designs and the priced bills of quantities formed part of the contract documents by virtue of an express provision in the contract. The foundation designs had been prepared on the assumption that the soil conditions were as indicated on the relevant borehole data provided by the corporation.

During the course of the work tufa was discovered in several areas of the site the presence of which required the foundations to be re-designed in those areas and additional works carried out. The contractor claimed that there had thereby been breach of an implied term or warranty by the corporation that the ground conditions would accord with the basis upon which the contractor had designed the foundations. They claimed they were entitled to be compensated by way of damages for breach of that term or warranty.

The corporation denied liability, maintaining that no such term or warranty could be implied but the court found in favour of the contractor.

2.1.9 The liability for inaccurate subsoil information provided by the employer at tender stage was an issue in *Co-Operative Insurance Society* v. *Henry Boot* (2002) in connection with a contract for the reconstruction of Lomond House in Glasgow.

In this case the employer, the Co-Operative Insurance Society, had commissioned Terra Tek Ltd to undertake a subsoil survey before tenders were sought. During the excavation work, water and soil flooded into the sub-basement excavations. It was alleged by Henry Boot that the problem stemmed from the inaccuracy of the ground water levels as shown in the subsoil survey. It argued that the method used for dewatering would have been adequate if the information on the subsoil survey had been accurate. The court was asked to decide as a preliminary issue whether the employer was responsible for the information contained in the Terra Tek Report. The contract bills under a heading 'Site Investigation' stated that the report had been issued to the contractor with the tender documents. In addition, drawing A303 expressly stated that the piling design and specification were based upon the Terra Tek Report and drawing 303A stated that 'for prevailing ground conditions refer to Site Investigation Report'. It looked to be game, set and match at this stage.

The court took a different view from that expressed on behalf of Henry Boot. Even though the drawings were referred to as contract documents and contained reference to the Terra Tek report, the judge felt that the report itself was not a contract document and did not form a part of the contract. The judge considered that clause 2.2.2.4 in the contract was more persuasive. This clause stated that the contractor is deemed to have inspected and examined the site and its surroundings and to have satisfied himself as to the nature of the site, including the ground and subsoil, before submitting his tender. The judge may have been legally correct but his decision defied the practicalities of preparing a tender and also common sense. Who would expect a contractor, provided at tender stage with a detailed subsoil survey, to visit the site of an existing building in a street in Glasgow and start digging up the ground to ensure that the information included in the Terra Tek Report was correct.

The conditions of contract used were the JCT conditions. If the conditions of contract had been ICE 7th Edition then the decision may have been different. Clause 11(2) indicates that the contractor is deemed to have inspected and examined the site and satisfied himself as to the form and nature of the subsoil but only in as far as is practicable and reasonable. This should be read with clause 11(3) which states that the contractor is deemed to have based his tender on his own inspection and all information whether obtained by him or made available by the employer.

2.1.10   It would seem that where an employer provides a site survey to contractors at tender stage they are entitled to rely upon it when calculating their tender price. If the survey proves to be inaccurate and as a result the contractor incurs additional costs he will usually be able to make out a good case for recovering those costs.

2.1.11   Many contracts are let however where the employer provides a subsoil survey but no specific reference is made to the employer taking responsibility. In many instances the specification will include a disclaimer. Despite silence in the contract or even a disclaimer, if the information is incorrect due to fraud or recklessness by the employer, architect or engineer, and the contractor suffers loss as a consequence, he may have a good case for recovering his additional costs.

Employers who specifically exclude liability for information provided or who limit the liability for incorrect information will have to show in accordance with the Unfair Contract Terms Act 1977 that the exclusion or limitation is reasonable. (See also 11.14) In any event exclusion clauses will not relieve employers from the results of their negligence unless liability for negligence is expressly excluded.

2.1.12 It may be in the employer's interests, when providing a subsoil survey, to make it clear that the information is intended to show the ground conditions which occur at the location of the boreholes only and on the dates on which they were taken. The contractor should be expressly informed not to assume that the conditions apply anywhere else on the site or at any later period.

2.1.13 It is worth noting that in the case of *Railtrack plc* v. *Pearl Maintenance Services Ltd* (1995) it was held that as the contract provided expressly for the contractor to ascertain the routes of services below ground he was liable for damage to services.

2.1.14 The decision in the USA case of *T.L. James and Co Inc* v. *Traylor Brothers Inc* (2002) drew attention to the pitfalls that contractors can encounter in not undertaking their own site investigations prior to submitting a tender.

Traylor submitted the lowest bid for the construction of a marine terminal, which included dredging, driving piles and constructing a deck and terminal building on the piles. The tender documents warned that the site had 'numerous steel and timber piles removed to the approximate existing mudline'. The documents also advised that a more detailed map was available in the archive rooms of the Port of New Orleans. Bidders were also instructed to undertake a pre-bid site investigation. Traylor neither looked at the map nor undertook more than a cursory site investigation.

James, Traylor's dredging subcontractor, encountered a large number of cut-off piles, and Traylor claimed compensation, arguing that the actual number and nature of the piles had not been disclosed and could not have been reasonably foreseen. The United States Court of Appeal, Fifth Circuit, rejected this. The real issue was whether the site conditions could have been anticipated, based on the information available at the time the bid was submitted. The court concluded that the conditions could have been anticipated, but were not because Traylor had failed to conduct an adequate site investigation.

SUMMARY

The ICE 6th and 7th Editions conditions make it clear that the contractor's tender is deemed to have been based upon his own inspection of the site and any information provided by the employer. With this in focus any additional costs resulting from unforeseen physical conditions will be recoverable.

In the absence of specific wording in the conditions the employer may wish to introduce a clause excluding liability should the site survey prove inaccurate. Whilst it is feasible to exclude liability if for any reason, including negligence, the information proves inaccurate, such an exclusion would have to be reasonable.

However the few cases dealing with this matter tend to lead to the conclusion that exclusion clauses of this nature are unlikely to find favour with the courts.

2.2   If, after tenders have been received, the employer decides not to proceed with the work, are there any circumstances under which the contractor/subcontractor can recover the costs associated with tendering?

2.2.1   It is not unusual for contractors or subcontractors to carry out work prior to a contract being let. Often this is done in contemplation of the contract being entered into to help the client, or to ensure a flying start or keep together key operatives. Sometimes, the contractor having made a start without the benefit of a contract, the employer decides not to proceed with the work. Contractors look to recover payment for their efforts and employers will usually deny any liability.

2.2.2   The case of *Regalian Properties plc* v. *London Docklands Development Corporation* (1991) dealt with this matter. Negotiations began in 1986 for the development of land for housing. A tender in the sum of £18.5m was submitted by Regalian for a licence to build when London Docklands obtained vacant possession. The offer was accepted 'subject to contract' and conditional upon detailed planning consent being obtained. Delays occurred in 1986 and 1987 due to London Docklands requesting new designs and detailed costings from Regalian. Due to a fall in property prices in 1988 the scheme became uneconomic, the contract was never concluded and the site never developed. Regalian claimed almost £3m which they had paid to their professional consultants in respect of the proposed development.

The court rejected the claim. The reasoning was that where negotiations intended to result in a contract are entered into on express terms that each party is free to withdraw from negotiations at any time, the costs of a party in preparing for the intended contract are incurred at its own risk and it is not entitled to recover them by way of restitution if for any reason no contract results. By the deliberate use of the words 'subject to contract' each party has accepted that if no contract was concluded any resultant loss should lie where it fell.

2.2.3   A different set of circumstances arose in the case of *Marston Construction Co Ltd* v. *Kigass Ltd* (1989).

A factory belonging to Kigass was destroyed in a fire in August 1986. Kigass thought that the proceeds from its insurance policy would cover the costs of rebuilding and accordingly invited tenders for a design and build contract for this work. Marston submitted a tender. It was believed by Kigass that the terms of its insurance policy required that the rebuilding work be performed as quickly as possible so Marston was invited to a meeting in December 1986 to discuss its tender. At this meeting it was made clear to Marston that no contract would be concluded until the insurance money was available, but both Marston and Kigass firmly believed that the money would be paid and that the contract would go ahead.

Marston received an assurance that it would be awarded the contract (subject to the insurance payment), but did not receive an assurance that it would be paid for the costs of the preparatory work.

Marston carried out the preparatory work but no contract was signed because the insurance money was insufficient to meet the costs of rebuilding.

It was established that no contract between the parties existed and there was no express request for preparatory work to be carried out. However the judge found in favour of the contractor on the basis that Kigass had expressly requested that a small amount of design work be carried out and that there was an implied request to carry out preparatory work in general.

2.2.4   A leading case on the subject matter is *William Lacey (Hounslow) Ltd* v. *Davis* (1957) where it was found that the contractor was entitled to be paid for preparatory work.

> The proper inference from the facts proved in this case is not that this work [i.e. the preparatory work] was done in the hope that this building might possibly be reconstructed and that the plaintiff company might obtain the contract, but that it was done under a mutual belief and understanding that this building was being reconstructed and that the plaintiff company was obtaining the contract.

2.2.5   Payment is due in restitution if the contractor, in the absence of a contract, is instructed to carry out work. In the case of *British Steel Corporation* v. *Cleveland Bridge & Engineering Co Ltd* (1981) the judge held:

> Both parties confidently expected a formal contract to eventuate. In those circumstances, to expedite performance under the contract, one requested the other to expedite the contract work, and the other complied with that request. If thereafter, as anticipated, a contract was entered into, the work done as requested will be treated as having been performed under that contract; if, contrary to that expectation, no contract is entered into, then the performance of the work is not referable to any contract the terms of which can be ascertained, and the law simply imposes an obligation on the party who made the request to pay a reasonable sum for such work as has been done pursuant to that request, such an obligation sounding in quasi contract, or, as we now say, restitution.

2.2.6   A situation may occur where work is undertaken on a speculative basis, for example preliminary design work, and it is agreed that payment will be made if the construction work on the project goes ahead.

In the case of *Dinkha Latchen* v. *General Mediterranean Holdings SA* (2003) an architect undertook preliminary design work in connection with a hotel and also a tennis club, both of which were in Tangiers, on the basis that no payment would be made until a building permit was obtained. The work began to escalate, with the client instructing the architect to do a great deal of design work and to visit Tangiers to engage local architects. It was the view of the court that there must come a point in the relationship when each party, had they addressed the question, would have recognised that there was no longer any intention that further work would be unremunerated.

It was held by the court that at some point between May 1994 and February 1995 the conduct of the parties was such as to give rise to an intention that any further work would be remunerated.

SUMMARY

The basic rule is that if a contractor carries out work prior to a contract being entered into he does so at his own risk. However, the court may order payment if it can be shown that the work was expressly requested or there was an implied requirement that the work be carried out.

2.3   Where a contractor/subcontractor submits a tender on request which is ignored, does he have any rights?

2.3.1   Contractors and subcontractors often fear that tenders they submit are not considered. The reasons can be varied. For example, the successful bidder of an earlier stage of the work is a almost certain to be awarded the work in a subsequent phase or the employer has a subsidiary who will do the work and tenders are invited merely to put pressure on the subsidiary to reduce the price.

2.3.2   In *Blackpool & Fylde Aero Club Ltd* v. *Blackpool Borough Council* (1990), the council put a concession to operate pleasure flights from the airport out to tender. Tenders were to be received by 12 noon on 17 March 1983. The letter box was supposed to be emptied by 12 noon each day, but was not. The club's tender was rejected as being late. The club maintained that the council had warranted that if a tender was returned by the deadline it would be considered and sought damages in contract for breach of that warranty, and in negligence for the breach of the duty that it claimed was owed to it.

It was held that the form of the invitation to tender was such that, provided an invitee submitted his tender by the deadline, he was entitled, under contract, to be sure that his tender would be considered with the others.

A tenderer whose offer is in the correct form and submitted in time is entitled, not as a matter of expectation but of contractual right, to be sure that his tender will be opened and considered with all other conforming tenders, or at least that his tender will be considered if others are.

A contractor or subcontractor whose properly submitted tender is not considered could levy a claim for damages which would normally be the abortive tender costs. Where the tender was submitted on a design and construct basis this could prove expensive. There may also be a possibility that the contractor or subcontractor could successfully frame a claim based upon loss of opportunity.

2.3.3   In the case of *Fairclough Building Ltd* v. *Borough Council of Port Talbot* (1992) a different set of circumstances arose.

The Borough Council of Port Talbot decided to have a new civic centre constructed and advertised for construction companies to apply for inclusion on the selective tendering list for the project.

Fairclough Building Ltd in March 1983 applied to be included on the list. Mr George was a construction director of Fairclough. His name appeared on Fairclough's letter of application. His wife, Mrs George, had been employed by the council since November 1982 as a senior assistant architect and the Borough Engineer was aware of the connection between Mrs George and Fairclough for some time, as she had disclosed this at the time that the council employed her.

Fairclough was invited to tender under the NJCC Two Stage Tendering Code of Procedure. Mrs George, now the Principal Architect, was to be involved in reviewing the tenders and wrote to the Borough Engineer reminding him of her connection with Fairclough.

The council considered the position and, having obtained counsel's opinion about how to proceed, decided to remove Fairclough from the select tender list for the project.

Fairclough brought proceedings for breach of contract. The Court of Appeal held that under the circumstances, the council had only two alternatives. One was to remove Fairclough from the tender list, and the other was to remove Mrs George altogether. In removing Fairclough from the list the council had acted reasonably. The council had no obligation to permit Fairclough to remain on the selected list and were not in breach of contract.

2.3.4   In giving its decision the Court of Appeal distinguished *Blackpool & Fylde Aero Club Ltd* v. *Blackpool Borough Council*. Blackpool and Fylde Aero Club did in fact submit a tender by posting it, by hand, in the council's letter box before the deadline. However, the council's staff failed to empty the letter box properly with the result that the tender was considered late, and rejected. In *Fairclough* the council had perhaps been in error in shortlisting Fairclough in the circumstances of the connection but had acted reasonably in removing them from the select tender list in the light of Mrs George's involvement.

2.3.5   The New Zealand case of *Pratt Contractors Ltd* v. *Transit New Zealand* (2003) dealt with a claim from a contractor whose tender was rejected despite being the lowest by over £1m. An appeal from a decision of the Court of Appeal in New Zealand was referred to the Privy Council in London. The reason for not awarding the contract to the lowest bidder was that the appointment of the successful contractor was based upon a 'weighted attribute method'. This system adopted a formula by which marks were given to quality attributes as well as price.

The Privy Council, in reaching a decision against the contractor, laid down a few ground rules to be applied when selecting a contractor. It was the duty of the employer to comply with any procedures set out in the tender enquiry. There was no obligation, however, for the employer to follow any of its own internal procedures. An obligation exists of good faith and fair dealings but this does not extend beyond a requirement for members to express views honestly held. This duty of fairness did not require the employer to appoint people who approached the task of tender selection with no preconceived views about the tenderers.

SUMMARY

A tenderer who submits a tender in the correct form has a contractual right to have his tender opened and considered but circumstances may arise in a particular case where conflicting duties make it reasonable for properly submitted tenders not to be considered.

2.4 Where a design and construct procurement method is employed, who is responsible for obtaining outline planning consent and full planning consent – the employer or contractor/subcontractor?

2.4.1 The use of a design and construct procurement method gives the employer a great deal of flexibility as to how detailed a scheme is developed before tenders are sought. One option is for an employer to go out to tender with no preconceived ideas as to the design he requires, where for example he owns a site with just a conception as to the type of building and floor area required.

At this stage, when the employer may not even have received planning consent merely expectations, a contractor could still be appointed. He would have a two stage brief, the first stage being to obtain outline planning consent. If this were not forthcoming the second stage would not proceed. An agreed fee could be arranged with the contractor for completing the first stage and there might be an arrangement whereby some or all of the fee was success related.

2.4.2 It is more usual, however for the employer to have obtained outline planning consent before appointing a contractor. Tenderers will be required to develop the internal and external details so that once the preferred solution and tender is accepted a submission can be made for full planning consent. If a two stage process is employed full planning consent may be sought after stage one tenders have been received but before the second stage price has been finally agreed. However employers are more inclined to pass responsibility for any changes which come about from obtaining full planning consent on to the contractor. Full planning consent is often not acquired until after the contract has been signed, leaving the contractor with the risk.

2.4.3 JCT With Contractor's Design clause 6.1.1.1 requires the employer's requirements to state specifically that they comply with statutory requirements which would include stating the extent to which planning consent had been obtained.

2.4.4 Clause 6.1.1.2 places responsibility upon the contractor to comply with all Acts of Parliament, regulations and by-laws. If planning consent, whether outline or fully detailed, has not been obtained by the employer clause 6.1.1.2 will place this responsibility upon the contractor.

SUMMARY

It is usual for the employer to obtain outline planning consent prior to going out to tender. Sometimes full planning consent is secured before the final price is agreed and often the employer secures both outline and full planning consent in advance of tender. There are differing methods employed where the contractor has to obtain both outline and full planning consent. Whichever system is employed the contract documents should make the position absolutely clear.

2.5 If an architect/engineer, acting as employer's agent in a Design and Construct contract approves the contractor's drawings and subsequently errors are found, will the architect/engineer have a liability?

2.5.1 A matter which needs to be addressed at the outset is whether the employer's agent should involve himself in approving contractor's drawings? Often the employer's agent is a quantity surveyor or other non design specialist without the appropriate expertise.

2.5.2 The standard forms of contract deal with the matter of the contractor submitting drawings to the Employer in several different ways.

- **JCT With Contractor's Design 1998 Edition:** Clause 5.3 requires the contractor to submit to the employer drawings which he intends using for the purposes of the works. No reference is made to any action to be taken by the employer or its agent on receipt of the drawings.
- **GC/Works 1/Design and Build (1998):** Condition 10A forbids the contractor from commencing work until the drawings have been submitted and they have been examined by the project manager who has either confirmed that he has no questions to raise in connection with the drawings, or that such questions have been raised and answered to his satisfaction.
- **ICE Design and Construct:** Clause 6(2)(a) requires the contractor to submit drawings to the employer's representative and not to commence work until the employer's representative consents thereto.

2.5.3 One standard form which makes reference to approval of contractor's drawings is MF/1 which refers in clause 16.1 to the engineer approving drawings.

   If, however, an employer's agent, whether he be 'architect or engineer, approves the contractor's drawings which are subsequently shown to include an error he may be liable. It would in the first instance be necessary to identify in the conditions of employment what his responsibilities were. Most consultancy agreements require the consultant to exercise due skill and care in carrying out his duties.

2.5.4 *George Fischer (GB) Ltd* v. *Multi Design Consultants Roofdec Ltd, Severfield Reece and Davis Langdon and Everest* (1998) is a case which, among other matters, examined the obligations of the employer's representative. George Fischer was the employer under an amended JCT With Contractor's Design. Multi Construction were main contractors and Davis Langdon and Everest both quantity surveyors and employer's representative. The project included the company's UK head office. From the outset the roof leaked and, despite some reduction of the problems following the taping over of the end lap joints, the leaking continued. The main contractor Multi Construction Ltd became insolvent and went into liquidation. A claim was made against Davis Langdon and Everest as a result of the roof leaks. The case against them was that under their contract with George Fischer they had an obligation to approve all working drawings. They were also, it was alleged, obliged to make visits to site to ensure that the work was being carried out in accordance with the drawings and specification. George Fischer claimed the problems with the roof would not have occurred had Davis Langdon and Everest carried out their duties as required by their contract. A further difficulty arose in that, under the terms of George Fischer's contract with Multi Construction, Davis Langdon and Everest were obliged to issue a certificate of practical completion which had the effect of releasing the bond.

Davis Langdon and Everest's defence was that they were not obliged to approve all working drawings. With regard to site visits they contended that access was unsafe which prevented them making site visits and even if such site visits were made the defective formation of the lap joints would not have been seen as workmen usually provide work of appropriate quality when being observed by the employer's representative. Their defence to the claim resulting from the issue of the certificate of practical completion was that in fact the certificate they issued was not one of practical completion but a substantial completion certificate accompanied by a two page document of incomplete work under a heading of reserved matters.

The judge was not impressed with Davis Langdon and Everest's defence. He considered that the contract made it clear that they were obliged to approve working drawings. The reasons given for not inspecting the work were dismissed. Davis Langdon and Everest's certificate he considered to be a certificate of practical completion as, apart from the final certificate, this was the only certificate of completion referred to in the contract.

The moral of the story is that those who work as employer's representatives should ensure that wording of their contracts with the employers is crystal clear as to the duties they are required to undertake and, for the avoidance of any doubt, equally clear as to the duties that are not required.

SUMMARY

If the employer's agent approves contractor's drawings he may have a liability. Most consultancy agreements require the consultant to exercise due skill and care in carrying out his duties. If, due to a failure to exercise due skill and care, an error remains unnoticed then the employer 's agent may have a liability. As with any claim based on allegations of negligence, the employer will have to demonstrate that the errors resulted in additional cost.

2.6   Does an employer have any liability for not sending a subsoil survey which is in his possession to tendering contractors, the absence of which leads a successful contractor to significantly underprice the risk of bad ground?

2.6.1   Some contracts specifically require employers to disclose information in their possession which is relevant. The FIDIC 1999 Edition, in clause 4.10 requires the employer to make available to the contractor all relevant data in the employer's possession on sub-surface and hydrological conditions. By way of contrast, the ICE 7th Edition, clause 11(1), provides for the contractor to take into account in the price only the information concerning the nature of ground and subsoil and hydrological conditions made available by the employer.

2.6.2   In the absence of specific obligations written into the contract, the law in the UK is vague on the matter. It has been argued that a failure to disclose information concerning ground conditions which may affect the contractor's tender price could amount to a negligent misstatement on the part of the employer. This line of argument was used in the Australian case of *Dillingham Construction Pty Ltd* v. *Downs* (1972), where it was recognised that the employer could owe the contractor a duty of care and this would include disclosure of relevant information. The decision went against the contractor as it was held that there was no reliance by the contractor on the employer providing the information.

2.6.3   There have been decisions of the Canadian courts which support the view that a duty of disclosure exists. In the case of *Quebec (Commission Hydro-electrique)* v. *Banque de Montreal* (1992), it was held that the law imposes a positive obligation to provide information in cases where one party is in a vulnerable position. In the case of *Opron Construction Co* v. *Alberta* (1994), the court took into account lack of time, the opportunities available for the tenderer to acquire the information, whether the information was indispensable and the degree of technicality of the data.

2.6.4   The Misrepresentation Act 1967 may come to the aid of contractors who have made no financial provision for bad ground conditions for which an allowance could have been made if the employer had issued information in its possession. It is possible for a liability to arise under the statute in respect of a failure to provide relevant information or where only a part of the information is provided. This was an issue in the decision of *Howard Marine and Dredging* v. *Ogden* (1978).

SUMMARY

The terms of the contract may require the employer to disclose information concerning ground conditions to the contractor. In the absence of a contractual obligation, the law in the UK is unclear.

In Canada, it has been held that the employer's obligation to disclose will depend upon time available, the opportunities for the tenderer to acquire the information, whether the information was indispensable and the degree of technicality of the data.

There has been no legal decision in the UK on this matter; however, in the event of such a case being heard, the court may consider that a failure to disclose amounts to a negligent misstatement, may involve a liability under the Misrepresentation Act 1967 or may follow the lead provided by the Canadian courts.

2.7   Where a tender enquiry requires tenders to remain open for acceptance for a specific period of time, can a contractor or subcontractor who has submitted a tender as required withdraw the tender before the period expires without incurring a financial liability?

2.7.1   It is a well established principle of English law that an offer can be withdrawn at any time up to it being unconditionally accepted. This rule applies even if the offeror undertakes to leave the offer open for a specified period of time. In the case of *Routledge* v. *Grant* (1828) the defendant offered to buy a house, giving the plaintiff six weeks to provide an answer. It was held that the offer could be withdrawn within the six-week period without incurring any liability. The situation would be different if consideration, e.g. a payment, were made to the offeror in return for keeping the offer open.

2.7.2   On international projects contractors are often required to provide a tender bond. If the tender is withdrawn before the period for acceptance has expired, the employer will be entitled to levy a claim against the bondsman.

2.7.3   The law in Hong Kong appears to differ from UK law. In the case of *City Polytechnic of Hong Kong* v. *Blue Cross (Asia Pacific) Insurance HCA* (1999) the plaintiff invited tenders from several insurance companies to provide health insurance for its staff. The tenders were required to remain open for acceptance for a period of three months. Blue Cross submitted the lowest tender but withdrew it before the three months period elapsed. The Polytechnic accepted another and more expensive insurer's tender and claimed the difference back from Blue Cross. At first instance, the claim was rejected. On appeal, however, the claim was successful on the grounds that there existed an implied contract that the tender would remain open for acceptance for a period of three months. Blue Cross was in breach of the implied contract and was obliged to pay damages.

Summary

Under English law an offer can be withdrawn at any time up to it being unconditionally accepted. This applies even if the offeror agrees to keep the offer open for a fixed period of time. Where some form of consideration, e.g. a payment, is provided to the offeror in return for keeping the offer open, it cannot be withdrawn until the period for acceptance expires. In Hong Kong it has been held that an undertaking to keep an offer open for acceptance for a given period of time represents an implied contract. A withdrawal of the offer before the period comes to an end constitutes a breach of contract.

2.8   **If a subcontractor submits a lump sum estimate to a contractor to carry out the subcontract work, and it is unconditionally accepted, can he later change the price on the basis that the lump sum was only an estimate?**

2.8.1   *Building Contract Dictionary* by Chappell, Marshall, Powell-Smith and Cavender, 3rd edition, published by Blackwell Publishing, provides two possible meanings to the word 'estimate':

- 'Colloquially and in the industry it means "probable cost" and is then a judged amount, approximate rather than precise'
- 'A contractor's estimate, in contrast, may, dependent upon its terms, amount to a firm offer, and if this is so, its acceptance by the employer will result in a binding contract'.

2.8.2   With two meanings attached to the word, using 'estimate' in the submission of an offer to carry out construction work can be ambiguous.

In the case of *Crowshaw* v. *Pritchard and Renwick* (1899) a contractor submitted a quotation for construction work in the following terms:

'ESTIMATE – Our estimate to carry out the sundry alterations to the above premises, according to the drawings and specifications, amounts to £1,230.'

2.8.3   The quotation was accepted but the contractor refused to proceed with the work. He contended that submitting an 'estimate' for carrying out the work was not intended to amount to an offer capable of acceptance. It was held by the court that the submission was a firm offer which had been accepted.

2.8.4   With this decision in mind, it is important for contractors and subcontractors to appreciate that, if the intention in submitting an estimate is that it is not intended to be a firm offer to carry out the work, this should be made clear. If, on the other hand, it is intended to be a firm offer but the price is subject to subsequent adjustment, then again this must be explained.

SUMMARY

The word estimate has two meanings. It may refer to a probable cost or approximate sum which can later be adjusted, or a firm price which is fixed. When submitting a quotation the term estimate should either not be used at all or qualified to provide its precise meaning.

3
Extensions of Time

3.1 Does a contractor or subcontractor lose entitlements to extensions of time if he fails to submit the appropriate notices and details required by the contract?

3.1.1 Most of the standard forms of main contract and subcontract require the contractor and subcontractor to give notice when delays occur to the progress or completion of the works. A question often asked is whether in the absence of notice the contractor or subcontractor loses his rights to have the completion date extended. In other words, is the service of notice a condition precedent to the right to an extension of time?

3.1.2 The matter was considered by the House of Lords in the case of *Bremer Handelsgesellschaft mbh* v. *Vanden Avenne-Izegem* (1978) which arose out of a dispute over the sale of soya bean meal. Lord Salmon, referring to how the rights of the parties were affected by the lack of a proper notice, had this to say:

> In the event of shipment proving impossible during the contract period, the second sentence of clause 21 requires the seller to advise the buyers without delay of the impossibility and the reasons for it. It has been argued by the buyers that this is a condition precedent to the sellers' rights under that clause. I do not accept this argument. Had it been a condition precedent, I should have expected the clause to state the precise time within which the notice was to be served and to have made plain by express language that unless the notice was served within the time the sellers would lose their rights under the clause.

3.1.3 From what Lord Salmon has said it seems clear that, for notice to be a condition precedent to a right for more time, the wording of the clause would need to be such that a failure to serve notice would result in loss of rights.

The situation of lack of notice was examined in the decision in *Stanley Hugh Leach* v. *London Borough of Merton* (1985) in relation to JCT 63 where Mr Justice Vinelott summarised the position as follows:

> The case for Merton is that the architect is under no duty to consider or form an opinion on the question whether completion of the works is likely to have been or has been delayed for any of the reasons set out in clause 23 unless and until the contractor has given notice of the cause of a delay that has become 'reasonably

apparent' or, as it has been put in argument, that the giving of notice by the contractor is a condition precedent which must be satisfied before there is any duty on the part of the architect to consider and form an opinion on these matters.

I think the answer to Merton's contention is to be found in a comparison of the circumstances in which a contractor is required to give notice on the one hand and the circumstances in which the architect is required to form an opinion on the other hand. The first part of clause 23 looks to a situation in which it is apparent to the contractor that the progress of the works is delayed, that is, to an event known to the contractor which has resulted or will inevitably result in delay. The second part looks to a situation in which the architect has formed an opinion that completion is likely to be or has been delayed beyond the date for completion. It is possible that the architect might know of events (in particular 'delay on the part of artists, tradesmen or others engaged by the Employer in executing work not forming part of this contract') which is likely to cause delay in completion but which has not caused an actual or prospective delay in the progress of the work which is apparent to the contractor. If the architect is of the opinion that because of an event falling within sub-paragraphs (a) to (k) progress of the work is likely to be delayed beyond the original or any substituted completion date he must estimate the delay and make an appropriate extension to the date for completion. He owes that duty not only to the contractor but also to the building owner. It is pointed out in a passage from *Keating on Building Contracts* 4th Edition at page 346, which is cited by the arbitrator, that if the architect wrongly assumes that a notice by the contractor is a condition precedent to the performance of the duty of the architect to form an opinion and take appropriate steps:

'... and in consequence refuses to perform such duties the Employer loses his right to liquidated damages. It may therefore be against the Employer's interests for an Architect not to consider a cause of delay of which late notice is given or of which he has knowledge despite lack of notice.'

3.1.4 The case of *City Inn* v. *Shepherd Construction* (2001) arose out of the construction of a hotel in Bristol. A dispute occurred concerning the contractor's entitlement to an extension of time. The contract was JCT 80 but with special amendments inserted at the behest of the employer. One of the amendments dealt with architect's instructions and required the contractor to serve a written notice with regard to an entitlement to an extension of time.

The contract stated that the architect could dispense with the contractor's obligations concerning written notice, but in the absence of such a waiver the special amendment stipulated that the contractor would not become entitled to any extension of time if he failed to serve the written notice.

Due to a failure on the part of the contractor to serve proper notice following the issue of an architect's instruction, the employer considered he had a right to deduct liquidated and ascertained damages at the rate of £30 000 per week stated in the contract. It was argued by the contractor that even though £30 000 per week represented a reasonable pre-estimate of the employer's loss, the removal of an entitlement to an extension of time due to a failure to serve notice was, in itself, a penalty.

The loss which the employer would incur as a result of a lack of written notice would still be the same. Therefore a failure by the contractor to serve a written notice had no financial effect upon the employer. The argument was not accepted by the court. If the contractor elects not to comply with the notice requirement, it can properly be regarded as a breach of contract. Should the contractor decide to serve a written notice then the employer provides an indemnity to the contractor against payment of liquidated and ascertained damages, but only if written notice is provided. A failure by the contractor to serve a written notice also prevents the architect from reviewing the instruction. For these reasons the court found against the contractor.

3.1.5   In *Maidenhead Electrical Services* v. *Johnson Controls* (1996) the terms of the contract laid down that any claim for an extension of time had to be made within ten days of the event for which the claim arises. It was held that a failure to comply with the notice provisions did not render a claim invalid.

3.1.6   The GC/Works/1 contract is somewhat out of line with the other standard forms of main contract in that it refers in condition 28 to the contractor's written notice being a condition precedent to a right to an extension of time unless otherwise directed by the Authority. However GC/Works/1 1998 form does not provide for the delay notice under condition 36 to be a condition precedent.

3.1.7   JCT 98 makes it clear under clause 25.3.3.1 that the architect's duties with regard to extending the completion date are not dependent upon service of notice by the contractor.

Clause 44 of the ICE 6th and 7th Editions is similarly worded.

3.1.8   The Engineering and Construction Contract (NEC) stipulates in clause 61.3 that the contractor must notify the project manager of a compensation event which has happened or which he expects to happen if:

- The contractor believes that the event is a compensation event
- It is less than two weeks since he became aware of the event
- The project manager has not notified the event to the contractor.

Clause 63.4 requires the project manager to assess the contractor's entitlement in the absence of proper notification.

3.1.9   The MF/1 and IChemE conditions make no reference to a delay notice being a condition precedent to an extension of time.

3.1.10   The interpretation of the various subcontracts runs in parallel with the main contracts. An exception is the CECA blue form subcontract for use with the ICE main contract. Clause 6(2) stipulates that it is a condition precedent to the subcontractor's rights to an extension of time for a notice to be served within 14 days of a delay first occurring for which the subcontractor considers himself entitled to extra time.

3.1.11   The Australian case *Turner Corporation Ltd (Receiver and Manager Appointed)* v. *Austotal Pty Ltd* (1998) dealt with the situation of a delay caused by the employer where the conditions of contract required a written delay notice as a condition precedent to an extension of time. The lack of notice lost the contractor the right to an extension of time. The judge stated:

> If the builder, having a right to claim an extension of time fails to do so, it cannot claim that the act of prevention which would have entitled it to an extension of time for practical completion resulted in its inability to complete by that time. A

party to a contract cannot rely on preventing conduct of the other party where it failed to exercise a contractual right which would have negated the effect of that preventing conduct.

3.1.12 Following the *Turner Corporation* case, the Australian courts seem to have undergone a change of opinion. The principle of prevention became relevant to extensions of time following the Australian case of *Gaymark Investments Pty Ltd* v. *Walter Construction Group Ltd* (1999). In basic terms the prevention principle states that one cannot benefit from one's own errors. Where an employer includes in a contract a clause which states that a written notice from the contractor is a condition precedent to the right to an extension, it could come into conflict with the prevention principles. A delay which is caused by the employer is obviously an employer's breach of contract. If the contract denies the contractor an entitlement to an extension of time due to a failure to serve a written notice, this could leave the employer with the benefit of being paid liquidated and ascertained damages which result from its own delays. In the *Gaymark* case, the contract required the contractor to serve a written notice within 14 days of the delay arising. The clause went on to stipulate that the contractor would not be entitled to an extension of time if it failed to serve a proper written notice. The court found in favour of the contractor even though there had been no written notice served as it considered it would be unmeritorious to award the employer liquidated and ascertained damages where the delays were of its own making.

3.1.13 Another Australian case, *Abigroup Contractors* v. *Peninsula Balmain* (2002) also involved the workings of an extension of time clause. The wording of the contract made the service of a written notice a condition precedent to the contractor's right to an extension of time. No notice was given but because the contract gave the superintendent power to grant an extension of time whether or not a notice had been served, he was wrong in not exercising the discretion and making an appropriate award.

3.1.14 A situation may arise where the contractor is delayed by the employer but loses rights to an extension of time due to the lack of a written notice which is expressed in the contract to be a condition precedent. The employer will become entitled to deduct liquidated damages. If the delay results in the contractor incurring additional cost he will still be entitled to recover them from the employer, either under the terms of the contract or as a common law damages claim: *Roberts* v. *Bury Improvement Commissioners* (1870). In this case the contractor in accordance with the express terms of the contract was entitled to an extension of time for employer's default which

> 'shall be deemed to be in full compensation and satisfaction for, or in respect of, any and every actual and probable loss or injury sustained or sustainable by the contractor'.

The architect was late in issuing drawings. It was held by the court that the wording of the contract did not prevent the contractor from claiming common law damages for breach.

SUMMARY

Where a contractor or subcontractor fails to serve a proper delay notice this will not result in the loss of rights to an extension of time unless the contract expressly states that the service of a notice is a condition precedent to such rights.

Following the Australian decision in *Gaymark Investments Pty Ltd* v. *Walter Construction Group Ltd* (1999) a requirement for a contractor to serve a written notice as a condition precedent to the right to an extension of time may fly in the face of the prevention principle. It was the court's view in this case that it would be unmeritorious to award the employer liquidated and ascertained damages where delays were of its own making. There is no evidence as yet of the UK courts following this decision. Under UK law, if a written notice is required by the terms of the contract as a condition precedent, the lack of it will result in the loss of a right to an extension of time even in the case of an employer's delay.

3.2 Are minutes of site meetings considered by the courts to be adequate notices of delay required by extension of time clauses?

3.2.1 Most standard forms of construction contracts provide in the extension of time clauses for the contractor and subcontractor to serve delay notices.

- **ICE 6th and 7th Editions – Clause 44(1)** The contractor shall, within 28 days after the cause of delay has arisen or as soon thereafter as is reasonable, deliver to the engineer full and detailed particulars in justification of any claim to an extension of time.
- **JCT 1998 – Clause 25.2.1.1** The contractor shall, if and when it becomes reasonably apparent that the progress of the works is delayed, forthwith give written notice to the architect.
- **GC/Works 1/1998 – Condition 36** Where the project manager receives notice requesting an extension of time from the contractor (which shall include grounds for the request) or where the project manager considers that a delay is likely, he must (as soon as possible or in any event within 42 days from the date of any notice) notify the contractor of his decision regarding an extension of time.
- **Engineering and Construction Contract (NEC) – Clause 61.3** The contractor is required to notify the project manager of an event which has happened or which he expects to happen if:
  - The contractor believes that the event is a compensation event
  - It is less than two weeks since he became aware of the event
  - The project manager has not notified the event to the contractor.

3.2.2 Before giving a definitive answer to the question it is necessary to decide whether a failure to serve a notice will lose the contractor or subcontractor his entitlement to an extension of time (see 3.1). If a formal notice is expressed as a condition precedent to the contractor's right to an extension of time, the question is often asked is whether minutes of site meetings constitute good notice.

3.2.3   In the Scottish case of *John L. Haley Ltd* v. *Dumfries & Galloway Regional Council* (1988) the court had to decide whether site meetings minutes constituted a good notice.

The contract in this matter was in the JCT 63 form under which the claimants undertook certain buildings works to a school with the contract period set at 78 weeks. This was overrun by 31 weeks with a six week extension of time granted. When further extensions were refused the matter was referred to arbitration. The claimants argued that they were entitled to an extension as the cause of delay fell within clauses 23(e), (f) and (h). The respondents maintained that the claimants were not entitled to an extension as they had not given written notice of the delay as required under clause 23.

The arbiter, following proof before answer, had granted a four week extension on the basis of site meeting minutes. At the respondent's instigation, the arbiter stated a case for the opinion of the Court of Session as to whether the minutes constituted a notice under clause 23.

The court held that the minutes did not constitute good notice. Unfortunately the claimants had conceded that notice was a condition precedent to an entitlement to an extension of time and lost their case.

3.2.4   Nevertheless the parties to a contract often agree at the outset that the obligation to submit written notices will be waived and delays will be recorded in the site meeting minutes instead. Where this occurs the employer would be estopped from denying the contractor an entitlement to an extension of time through lack of written notice.

SUMMARY

Whether site meetings minutes constitute a good delay notice will depend upon the precise wording of the contract. It would seem, however, following the Scottish decision of *John L. Haley Ltd* v. *Dumfries & Galloway Regional Council* (1988) that in the case of the majority of the standard forms the site meeting minutes will not constitute good notice unless the parties specifically amend the contract in this respect.

3.3   What is meant by a contractor or subcontractor having to 'use constantly his best endeavours to prevent delay'?

3.3.1   Many contracts require a contractor or subcontractor to use constantly his best endeavours to prevent delay. For example JCT 98 clause 25.3.4 states:

> 'the Contractor shall use constantly his best endeavours to prevent delay in the progress of the Works...'

Best endeavours means that all steps to achieve the objective must be taken. *Keating on Building Contracts* 5th edition at page 575 has this to say with regard to the wording as it appears in the JCT forms of contract:

> 'This proviso is an important qualification of the right to an extension of time. Thus, for example, in some cases it might be the contractor's duty to reprogramme the works either to prevent or to reduce delays. How far the contractor must take the other steps depends upon the circumstances of each case, but it is thought that the proviso does not contemplate the expenditure of substantial sums of money.'

3.3.2 The wording of I Chem E clause 14.3 is a little different where it states that the contractor shall at all times use his best endeavours to minimise any delay in the performance of his obligations under the contract.

3.3.3 GC/Works/1 1998 in condition 36(6) states that the contractor must endeavour to prevent delays and to minimise unavoidable delays.

3.3.4 In the case of *IBM UK Ltd* v. *Rockware Glass Ltd* (1980), Rockware agreed to sell IBM some land for development, and the sale was conditional upon planning permission being obtained with a further proviso that IBM 'will make an application for planning permission and use its best endeavours to obtain the same'. The local authority refused planning permission. IBM did not appeal against that decision to the Secretary of State. The parties disagreed on whether, by not appealing, IBM had failed to use its best endeavours to obtain planning permission.

The project was a substantial one in which the purchase price of the land alone was £6 250 000. It was accepted that making an appeal to the Secretary of State would cost a significant amount of money.

The court said that, taking into account the background to the matter and the amount of money involved, it was not likely that the parties would have considered a refusal of planning permission at a local level to be the end of the matter, but that they must have had in mind the prospect of an appeal to the Secretary of State. The test of best endeavours which was approved was that the purchasers of the land were

> bound to take all those steps in their power which are capable of producing the desired results, namely the obtaining of planning permission, being steps which a prudent, determined and reasonable owner, acting in his own interests and desiring to achieve that result, would take.

It was expressly stated that the criterion was not that of someone who was under a contractual obligation, but someone who was considering his own interests.

Whilst it seems clear a contractor or subcontractor may be required to expend some money to meet the obligation to use constantly his best endeavours to prevent delay, the intention is not to expend large sums, particularly where the delay has been caused by the engineer or architect.

3.3.5 In the case of *Victor Stanley Hawkins* v. *Pender Bros Pty Queensland* (1994) it was held that the term 'best endeavours' should be construed objectively. The test as to whether it had been fulfilled would be that of prudence and reasonableness.

3.3.6 There is a difference between the meaning of best endeavours and reasonable endeavours. Reasonable endeavours involves considering all relevant circumstances including the commercial and financial aspects, but is not so high a test.

3.3.7 Two cases have involved the court in having to decide the meaning of best endeavours. In *Midland Land Reclamation Ltd* v. *Warren Energy Ltd* (1997) the judge in deciding the case said:

> I reject the submission made on behalf of the defendant that a best endeavours obligation is the next best thing to an absolute obligation or guarantee.

In *Terrell* v. *Maby Todd & Co* (1952) the judge held that a best endeavours obligation only required a party to do what was commercially practicable and what it could reasonably do in the circumstances.

3.3.8   In an article in *Building* (10 September 1999) Neil White explained:

> 'A best endeavours clause means that you do what a reasonable person would to achieve an objective – it is not a guarantee it may be overruled by conflicting obligations and it doesn't apply to intangible outcomes such as an agreement.'

In the final analysis a contractor will be expected to do what is commercially practicable and what it could reasonably do in the circumstances.

SUMMARY

Where a contract requires the contractor to use his best endeavours to prevent delay he is expected to keep the effect of any matters which could cause delay down to a minimum or to eliminate them if possible. If the delay is the contractor's responsibility he may consider it financially more advantageous in the absence of an obligation to use best endeavours, to allow the work to overrun the contract period and to pay liquidated damages. This will be particularly relevant if the liquidated damages figure is modest.

   If the delay is the responsibility of the architect/engineer or employer the contractor is not required to expend substantial amounts of his own money to reduce the delay.

3.4   If the architect/engineer issues a variation after the extended completion date but before practical completion, can an extension of time be granted or will liquidated damages become unenforceable?

3.4.1   Architects and engineers should issue variations for extra work at times appropriate to the progress of the works. Often, where delays have occurred, variations are issued after the date has passed when work should have been completed. Contractors frequently argue that, due to the timing of the variation, an extension of time should be granted up to the date the variation was issued plus adequate time to carry out the extra work. A contrary argument is often made that the issue of a variation after the completion date in the contract has passed does not rank for an extension of time and leaves time at large requiring the contractor to finish in a reasonable time.

3.4.2   The case, *Balfour Beatty Building Ltd* v. *Chestermount Properties Ltd* (1993) heard before Mr Justice Colman of the commercial court, arose out of an appeal against an award of Christopher Willis, a well known and respected arbitrator, and deals with the subject matter in question.

3.4.3 The works employing JCT 80 comprised the construction of the shell and core of an office block. Work commenced in September 1987, the completion date being 17 April 1989, later extended to 9 May 1989. As work was not completed by this date a certificate of non-completion was issued by the architect under clause 24.1. By January 1990 the work had still not been completed.

During the period 12 February 1990 to 12 July 1990 the architect issued instructions for the carrying out of fit out works as a variation to the contract. Practical completion of the shell and core was achieved on 12 October 1990 with the fit out works not finished until 25 February 1991. The architect issued two extensions of time to give a revised completion date of 24 November 1989. The variations with regard to the fit out works were issued after the revised completion date but prior to practical completion, during a period of default. The architect then revised the non-completion certificate to reflect the extended completion date.

3.4.4 The contractor argued that the effect of the issue of variations during a period of culpable delay was to render time at large, leaving the contractor to complete within a reasonable time. This being the case, the employer would lose his rights to levy liquidated damages.

Alternatively, the contractor contended that the architect should have granted an extension of time on a gross basis. In this case it was argued that the fit out work should have taken 54 weeks, this period to be added to 12 February 1990, when the fit out variation was issued.

It was the employer's contention that the correct approach should be a net extension of time, that is to say, one which calculated the revised completion date by taking the date currently fixed for completion and adding to it the 18 weeks that the architect considered to be fair and reasonable for the fit out work.

The main plank in support of the contractor's argument was that if the net method was adopted the extended completion date would expire before the variation giving rise to the extension had been instructed, which was logically and physically impossible. If the contractor's line were followed it would provide him with a windfall which swept up his delays. While recognising this, the contractor considered the problem resulted from the employer's own voluntary conduct in requiring a variation during a period of culpable delay.

3.4.5 Mr Justice Colman did not agree. He found in favour of the employer for the following reasons:

- When the architect reviews extensions of time under clause 25.3.3.2 following practical completion he is entitled to reduce the extended contract period to take account of omissions. These may have been issued during a period of culpable delay. It would, therefore, be illogical for the architect to have to deal with additions differently to the way he deals with omissions.
- The objective of clause 25.3.1 is for the architect to assess whether any of

the relevant events have caused a delay and if so by how much. He must then apply the result of his assessment to give a revised completion date. It would need clear words in the contract to allow the architect to depart from a requirement to postpone the completion date by the period of delay caused by the relevant event.

Mr Justice Colman finally concluded:

> In the case of a variation which increases the works, the fair and reasonable adjustment required to be made to the period for completion may involve movement of this completion date to a point in time which may fall before the issue of the variation instruction.

3.4.6 This decision is unlikely to apply to ICE 6th and 7th Editions where under clause 47(6) liquidated damages are suspended during a period of delay resulting from variations, a clause 12 situation, or any other delaying event outside the control of the contractor.

SUMMARY

Where an architect/engineer issues a variation after the contract completion date but before practical completion, it is appropriate where resultant delays occur for an extension of time to be granted. Such extension of time will be calculated by extending the completion date by the net period of delay. This is unlikely to apply to ICE 6th and 7th Editions which provide for the suspension of liquidated damages during a period of delay caused by variations and the like.

3.5   If work is delayed due to two or more competing causes of delay, one of which is the responsibility of the contractor/ subcontractor or a neutral event and the other as a result of some fault of the architect, engineer or employer, is there an entitlement to an extension of time or loss and expense?

3.5.1 Delays can be excusable delays which may be either due to fault by the employer or his agents, such as late access or delay in issuing drawings or delays due to a neutral event such as excessively adverse weather or force majeure. Inexcusable delays are those which are due to fault on the part of the contractor or his agents.

3.5.2 Arguments as to a contractor or subcontractor's entitlements where two competing causes of delay occur which affect the completion date have never been easy to resolve. *Keating on Building Contracts* 5th edition has offered assistance by suggesting a number of approaches, as at p. 193 which includes:

- **The Devlin approach:** 'If a breach of contract is one of two causes of a loss, both causes co-operating and both of approximately equal efficacy, the breach is sufficient to carry judgment for the loss.'

This would apply where, for example, there were two competing causes of delay which entitled a contractor to an extension of time, one a neutral event such as excessively adverse weather and the other being a breach such as late issue of instructions by the architect. Following the Devlin approach the contractor would be entitled to extra time and loss and expense due to the late issue of instructions.

- **The dominant cause approach:** 'If there are two causes, one the con-tractual responsibility of the defendant and the other the contractual responsibility of the plaintiff, the plaintiff succeeds if he establishes that the cause for which the defendant is responsible is the effective, dominant cause. Which cause is dominant is a question of fact, which is not solved by the mere point of order in time, but is to be decided by applying common sense standards.'

- **The burden of proof approach:** 'If part of the damages is shown to be due to a breach of contract by the plaintiff, the claimant must show how much of the damage is caused otherwise than by his breach of contract, failing which he can recover nominal damages only.'

  An example would be a delay caused by the contractor having to correct defective work running at the same time as a delay caused by the employer. Little in the way of extra cost would be recoverable as it would be difficult for the contractor to demonstrate that his losses were due to the employer's actions and not his own.

3.5.3   The dominant cause of delay theory was rejected by the court in the case of *H Fairweather & Co Ltd* v. *London Borough of Wandsworth* (1987). H Fairweather and Co Ltd were the main contractors for the erection of 478 dwellings for the London Borough of Wandsworth employing JCT 63 conditions. Long delays occurred and liability for those delays was referred to arbitration.

With regard to the delays the architect granted an extension of 81 weeks under condition 23(d) by reason of strikes and combination of workmen. The quantum of extension was not challenged but Fairweather contended before the arbitrator that 18 of those 81 weeks should be reallocated under clause 23(e) or (f). The reasoning behind the contention was that only if there was such a reallocation could Fairweather ever recover direct loss and expense under clause 11(6) in respect of those weeks reallocated to clause 23(e) or clause 24(1)(a) in respect of those weeks reallocated to clause 23(f).

The arbitrator's reasoning is to be found in sections 6.11 and 6.12 of his interim award:

> 6.11   It is possible to envisage circumstances where an event occurs on site which causes delay to the completion of the works and which could be ascribed to more than one of the eleven specified reasons but there is no mechanism in the condi-tions for allocating an extension between different heads so the extension must be granted in respect of the dominant reason.

> 6.12.3   I accept the respondent's contention that, faced with the events of this contract, nobody would say that the delays which occurred in 1978 and 1979 were caused by reason of the Architect's instructions given in 1975 to 1977. I hold that the dominant cause of the delay was the strikes and combination of workmen and

accordingly the Architect was correct in granting his extension under condition 23(d).

In 6.14 he said:

> For the sake of clarity I declare that this extension does not carry with it any right to claim direct loss and/or expense.

The arbitrator's award was the subject of an appeal. The judge in the case disagreed with the arbitrator's ruling that the extension of time should relate to the dominant cause of delay. He said in his judgment:

> 'Dominant' has a number of meanings: 'Ruling, prevailing, most influential'. On the assumption that condition 23 is not solely concerned with liquidated or ascertained damages but also triggers and conditions a right for a contractor to recover direct loss and expense where applicable under condition 24 then an architect and in his turn an arbitrator has the task of allocating, when the facts require it, the extension of time to the various heads. I do not consider that the dominant test is correct.

This decision places doubt upon *Keating's* 'Dominant Cause' theory.

3.5.4   The dominant cause approach, however, received the following support:

(1) *Leyland Shipping* v. *Norwich Union* (1918)

> 'Which cause is dominant is a question of fact which is not solved by mere point of order in time, but is to be decided applying common sense standards.'

(2) *Keating on Building Contracts*, 6th edition, as at p. 212

> 'The burden of proof should require the plaintiff to establish that the loss resulted from the cause on which he relies and that he will do so if he establishes that it was the dominant cause. He should not be required additionally to disprove lesser alternative causes altogether.'

3.5.5   There is another rule which is applicable to concurrent delays. Where an employer delays the contractor he will not be entitled to deduct liquidated damages even though the contractor is also in default. (*Wells* v. *Army & Navy Co-operative Society* (1903).

With this in mind Keith Pickavance in his book *Delay and Disruption in Construction Contracts* at p. 352 states:

> 'Lastly, and this is a legal conceptual problem, the rules which apply to recovery of actual damages for delay are not the same rules that apply to the relief of liquidated damages for delay. If C's progress on the critical path has been interfered with by D's act of prevention, then C must be given sufficient time to accommodate the effects of that and be relieved for LADs [liquidated and ascertained damages] for a commensurate period.
>
> On the other hand if, during the period of disruption to progress or prolongation for which an EOT [extension of time] has been granted, the predominant cause of C's loss and expense is disruption, or prolongation caused by a neutral event or his own malfeasance (for which he bears the risk), then he will not be able to recover damages for the compensable event unless he can separate those costs flowing from the compensable event from those costs which are at his own risk.'

In other words if two delays are running in parallel one cause being the contractor's default the other a breach by the employer, an extension of time should be awarded to the contractor but no monetary reimbursement.

3.5.6　The courts in the USA have also addressed this problem and applied the legal maxim that a party cannot benefit from its own errors. An employer who deducts liquidated damages during an overrun period when the delay is being caused by both late issue of information and correcting defective work running concurrently could fall into this category. The USA courts have taken the line that where this type of situation arises the employer will not be entitled to deduct liquidated damages and for the same reason the contractor will not be entitled to payment of additional cost.

3.5.7　A simplistic approach sometimes taken is the 'first past the post' approach. This adopts the logic that where delays are running in parallel the cause of delay which occurs first in terms of time will be used for adjustment of the contract period. Other causes of delay will be ignored unless they affect the completion date and continue on after the first cause has ceased to have any delaying affect. In this case only the latter part of that second cause of delay will be relevant to the calculation of an extension of time.

For example delays may run in parallel due to the late issue of drawings and inclement weather. If the late issue of drawings causes a delay commencing on 1 February and inclement weather makes work impossible from 14 February the late issue of drawings is the 'first past the post' and will take precedence over inclement weather until the drawings are issued. If drawings are issued on 21 February but the inclement weather continues until 28 February the contract completion date will be adjusted in respect of weather for the latter period.

The Australian Building Works Contract, JCC-E 1994 published by the Joint Contracts Committee, under clause 10.11 'Predominant Causes of Delay' applies the 'first past the post' approach to delays caused by the proprietor's default.

3.5.8　It has been suggested that when delays are concurrent some form of apportionment may be appropriate. The matter of allocating or apportionment was referred to in *Keating on Building Contracts* 6th edition, p. 213:

> 'Where the loss or damage suffered by the plaintiff results partly from his own conduct and partly from the defendant's breach of contract, it is correct in principle for the damages to be apportioned.'

This seems to be at variance with Keating's own burden of proof approach referred to in section 3.5.2.

In the case of *Tennant Radiant Heat Ltd* v. *Warrington Development Corporation* (1987) a warehouse roof collapsed due to an accumulation of rainwater because of blocked outlets. The landlord had a liability for damages and the tenant the responsibility for repair, both of which could have caused the collapse. The court, under the circumstances, apportioned the loss between both of the matters which could have contributed to the cause of the collapse. Lord Justice Dillon explained the rationale as follows:

'The problem which this court faces, on the claim and counterclaim alike, is in my judgment a problem of causation of damages. On the claim, the question is how far the damage to its goods which the lessee has suffered was caused by the corporation's negligence notwithstanding the lessee's own breach of covenant. On the counterclaim, the question is how far the damage to the corporation's building which the corporation has suffered was caused by the lessee's breach of covenant, notwithstanding the corporation's own negligence. The effect is that on each question, apportionment is permissible.'

3.5.9 In the case of *John Doyle Ltd* v. *Laing Management (Scotland) Ltd* (2004) the court had another opportunity of giving its opinions concerning concurrent delays. It seemed to favour the dominant cause approach. However, if the dominant cause approach could not be applied, then it considered apportionment may be appropriate.

The court, in explaining the significance of dominant cause and apportionment, stated as follows:

'[14] ... it may be possible to identify a causal link between particular events for which the employer is responsible and individual items of loss. On occasion that may be possible where it can be established that a group of events for which the employer is responsible are causally linked with a group of heads of loss, provided that the loss has no other significant cause. In determining what is a significant cause, the 'dominant cause' approach described in the following paragraph is of relevance. Determining a causal link between particular events and particular heads of loss may be of particular importance where the loss results from mere delay, as against disruption; in cases of mere delay such losses as the need to maintain the site establishment for an extended time can often readily be attributed to particular events, such as the late provision of information or design changes...
[15] ... the question of causation must be treated by the application of common sense to the logical principles of causation... In this connection, it is frequently possible to say that an item of loss has been caused by a particular event notwithstanding that other events played a part in its occurrence. In such cases, if an event or events for which the employer is responsible can be described as the dominant cause of an item of loss, that will be sufficient to establish liability, notwithstanding the existence of other causes that are to some degree at least concurrent...
[16] ... even if it cannot be said that events for which the employer is responsible are the dominant cause of the loss, it may be possible to apportion the loss between the cause for which the employer is responsible and other causes. In such a case it is obviously necessary that the event or events for which the employer is responsible should be a material cause of the loss. Provided that condition is met, however, we are of the opinion that apportionment of loss between the different causes is possible in an appropriate case. Such a procedure may be appropriate in the case where the causes of the loss are truly concurrent, in the sense that both operate together at the same time to produce a single consequence.'

3.5.10 A 'but for' test has been developed which can be advantageous to contractors. The argument often runs that but for the architect's instruction the delay would not have occurred. In the Australian case of *Quinn* v. *Burch Brooks (Builders) Ltd* (1966), auditors failed to identify that a company was in a substantial loss position. It was argued that but for the auditor's errors the loss making company would have ceased trading and subsequent trading losses would therefore have been avoided.

In the UK case of *Turner Page Music Ltd* v. *Torres Design Associates Ltd* (1997) an action was brought against an architect for negligence when there was a severe financial overrun on a construction project. The employer subsequently sold the building and alleged that but for the architect's negligence the sale would not have been necessary.

The courts in both cases rejected the 'but for' argument.

3.5.11 Judge Dyson, in the case of *Henry Boot Construction (UK) Ltd* v. *Malmaison Hotel (Manchester) Ltd* (1999), was even more generous to contractors when, without dissent, he included the following matters which had surprisingly been agreed by both parties:

> '... it was agreed that if there are two concurrent causes of delay, one of which is a relevant event, and the other is not, then the contractor is entitled to an extension of time for the period of delays caused by the relevant event notwithstanding the concurrent effect of the other event. Thus, to take a simple example, if no work is possible on a site for a week not only because of exceptionally inclement weather (a relevant event), but also because the contractor has a shortage of labour (not a relevant event), and if the failure to work during that week is likely to delay the work beyond the completion date by one week, then if he considers it fair and reasonable to do so, the architect is required to grant an extension of time of one week. He cannot refuse to do so on the grounds that the delay would have occurred in any event by reason of the shortage of labour.'

3.5.12 The Society of Construction Law Delay and Disruption Protocol 2002 presents a view which is similar to the burden of proof approach referred to in 3.5.2, when it states:

> '1.10.4 Where an Employer Risk Event and a Contractor Risk Event have concurrent effect, the Contractor may not recover compensation in respect of the Employer Risk Event unless it can separate the loss and/or expense that flows from the Employer Risk Event for that which flows from the Contractor Risk Event. If it would have incurred the additional costs in any event as a result of Contractor Delays, the Contractor will not be entitled to recover those additional costs. In most cases this will mean that the Contractor will be entitled to compensation only for any period by which the Employer Delay exceeds the duration of the Contractor Delay.
>
> 1.10.5 The loss and/or expense flowing from an Employer Delay cannot usually be distinguished from that flowing from Contractor Delay without the following:
>
> > 1.10.5.1 an as-planned programme showing how the Contractor reasonably intended to carry out the work and the as-planned critical path;
> >
> > 1.10.5.2 an as-built programme demonstrating the work and sequence actually carried out and the as-built critical path;
> >
> > 1.10.5.3 the identification of activities and periods of time that were not part of the original scope;
> >
> > 1.10.5.4 the identification of those activities and periods of time that were not part of the original scope and that are also at the Contractor's risk as to cost; and

1.10.5.5 the identification of costs attributable to the two preceding sub-sections.'

SUMMARY

There is no hard and fast rule concerning which delay takes precedence where a number of delays affect the completion date. Each case has to be judged on its own merits.

3.6   Is a notice which does not quote contract clause numbers adequate or does the contractor/subcontractor risk losing his contractual rights where these references are omitted?

3.6.1   It is essential if contractors and subcontractors are to avoid the risk of losing their rights, to ensure that such written notices as are required by the contract are served in a correct and timeous manner. The wording of the clause with regard to what details must be included in the notice may be sufficiently clear to avoid uncertainty. However, frequently it is a little vague as to what is required in the way of notice.

Often disputes arise where a contractor or subcontractor serves what he considers to be an adequate notice to obtain a right under a particular clause. Some time later, when denying the rights, an employer may wish to plead inadequate notice. This can create serious problems for contractors and subcontractors if the notice is to be served within a fixed timescale and a purported notice proves to be defective. (See *Monmouthshire County Council* v. *Costello & Kemple Ltd* (1965), *Blackpool Borough Council* v. *F. Parkinson Ltd* (1991) and *Rees & Kirby Ltd* v. *Swansea City Council* (1983).)

3.6.2   In the case of *London Borough of Merton* v. *Stanley Hugh Leach Ltd* (1985), the court had to decide what constituted good notice under JCT 63. Mr Justice Vinelott had this to say:

> But in considering whether the contractor has acted reasonably and with reasonable expedition it must be borne in mind that the architect is not a stranger to the work and may in some cases have a very detailed knowledge of the progress of the work and of the contractor's planning. Moreover, it is always open to the architect to call for further information either before or in the course of investigating a claim. It is possible to imagine circumstances where the briefest and most uninformative notification of a claim would suffice: a case, for instance, where the architect was well aware of the contractor's plans and of a delay in progress caused by a requirement that works be opened up for inspection but where a dispute whether the contractor had suffered direct loss or expense in consequence of the delay had already emerged. In such case the contractor might give a purely formal notice solely in order to ensure that the issue would in due course be determined by an arbitrator when the discretion would be exercised by the arbitrator in the place of the architect.

3.6.3   *Keating on Building Contracts,* 4th edition gives helpful guidance on notice and the contents of the application:

'A consideration of the effect of the *Minter* case raises some questions about what an application should contain in order to be valid under clause 11(6). No form is required, but it must, it is submitted, be expressed in such a way, or made in such circumstances, as to show that the Architect is being asked to form the opinion referred to in the sub-clause and identify the variation or provisional sum work relied on. It is thought that great particularity is not contemplated by the sub-clause. The Architect must know the variation of work relied on; he must also either be told in the application, or must be taken to know from his knowledge of the circumstances, sufficient to enable him to form the opinion that the contractor has been involved in direct loss and/or expense for which he would not be reimbursed under sub-clause (4). He does not have to have at the stage of forming his opinion sufficient details to ascertain the amount; it is sufficient if he has enough to form the view that there must be some loss.'

SUMMARY

There is no legal requirement for reference to be given in a contractual notice to the clause number under which it is given. In the interests of clarity, however, the reference should be provided.

3.7 **Where delays occur to the main contract works due to the time taken to correct latent defects in a nominated subcontractor's work after the subcontract works have been completed, does this give the main contractor the right to an extension of time if the contract makes provision for extensions of time due to delays on the part of nominated subcontractors?**

3.7.1 JCT 98 provides under clause 25.4.7 for an extension of time due to

'delay on the part of Nominated Sub-contractors or Nominated Suppliers which the Contractor has taken all practical steps to avoid or reduce'.

If a nominated subcontractor completes its installation but has subsequently to return to site to correct a latent defect and as a result the main contractor suffers a delay to completion, will there be an entitlement to an extension of time under this clause?

3.7.2 The House of Lords had to decide this point in the case of *Westminster City Council* v. *J Jarvis & Sons Ltd* (1970). A JCT 63 form of contract was used, the provisions of clause 23(g) being similar to clause 25.4.7 of JCT 98.

J Jarvis & Sons entered into a contract with the City of Westminster for the construction of a multi-storey car park. The date for completion was 15 January 1968.

Peter Lind & Co Ltd were the nominated sub-contractors for piling work. The date for completion of their work was 21 June 1966 and they purported to have completed their works by that date, withdrawing from the site. A month later, on 21 July 1966, one of the piles was damaged. It was then discovered that many of the piles were defective. The nominated subcontractor installed replacement piles but the remedial work was not completed until 29 September 1966. As a result of the necessity for such remedial work the works under the main contract were delayed.

3.7.3 It was held that, on a proper interpretation of clause 23(g), delay on the part of a nominated subcontractor only occurred if, by the date for completion in the subcontract, the subcontractor had failed to achieve such completion of his work that he could not hand it over to the contractor.

If the subcontract works are apparently completed and taken over by the main contractor delays caused by the subcontractor returning to site to correct latent defects will not rank for an extension of time.

SUMMARY

If a contract such as JCT 98 provides for extensions of time due to delays on the part of a nominated subcontractor, extensions of time will not become due where delays have occurred to the main contract completion date as a result of a subcontractor returning to site to make good latent defects.

3.8 Can an architect/engineer grant an extension of time after the date for completion has passed?

3.8.1 It is desirable for extensions of time to be granted at such times that contractors always know in advance the date for completion to which they are working. Some contracts will be precise as to when an architect/engineer must deal with extensions of time. JCT 98 for example under clause 25.3.1 states that the architect shall:

> 'if reasonably practicable having regard to the sufficiency of the aforesaid notice, particulars and estimates, fix such new Completion Date not later than 12 weeks from receipt of the notice and of reasonably sufficient particulars and estimates, or, where the period between receipt thereof and the Completion Date is less than 12 weeks, not later than the completion date.'

The ICE 7th Edition states under clause 44(5):

> 'The Engineer shall within 28 days [14 days in the 6th Edition] of the issue of the Certificate of Substantial Completion for the Works or any Section thereof review all the circumstances of the kind referred to in sub-clause (1) of this Clause and shall finally determine and certify to the Contractor with a copy to the Employer the overall extension of time (if any) to which he considers the Contractor entitled ...'

GC/Works/1 1998 condition 36 requires the project manager to grant extensions of time

> 'as soon as possible and in any event within 42 days from the date any such notice is received ...'

3.8.2 The question of the timing of an extension of time award has been discussed in the following legal cases where the contracts were not precise as to the timing of an extension of time award.

3.8.3   In *Miller* v. *London County Council* (1934) the express wording of the contract provided:

> 'It shall be lawful for the Engineer, if he thinks fit, to grant from time to time, and at any time or times, by writing under his hand such extension of time for completion of the work and that either prospectively or retrospectively and to assign such other time or times for completion as to him may seem reasonable.'

It was held that the words 'either prospectively or retrospectively' did not give the engineer power to fix a new date for completion after the completion of the works.

3.8.4   In *Amalgamated Building Contractors Ltd* v. *Waltham Holy Cross Urban District Council* (1952) the wording in the contract which was the then current RIBA contract provided in clause 18 that 'the Architect shall make a fair and reasonable extension of time for completion of the works'.

Lord Denning, with regard to the time within which the architect would be required to make a decision, had this to say:

> The contractors say that the words in clause 18 mean that the architect must give the contractors a date at which they can aim in the future, and that he cannot give a date which has passed, I do not agree with this contention.

Lord Denning distinguished the decision in *Miller* v. *LCC* in the following terms:

> These practical illustrations show that the parties must have intended that the architect should be able to give a certificate which is retrospective, even after the works are completed ... *Miller* v. *London County Council* (1934) is distinguishable. I regard that case as turning on the very special wording of the clause which enable the engineer 'to assign such other time or times for completion as to him may seem reasonable'. Those words, as Mr Justice du Parcq said, were not apt to refer to the fixing of a new date for completion *ex post facto*. I would also observe that on principle there is a distinction between cases where the cause of delay is due to some act or default of the building owner, such as not giving possession of the site in due time, or ordering extras, or something of that kind. When such things happen the contract time may well cease to bind the contractors, because the building owner cannot insist on a condition if it is his own fault that the condition has not been fulfilled.

SUMMARY

The timing of the architect/engineer's decisions concerning the granting of an extension of time may be stipulated in the contract. This being the case the architect/engineer will be required to comply with the requirements.

If no time is laid down there seems no impediment to the Architect or Engineer making a decision after the date for completion has passed.

3.9   When an architect/engineer is considering a contractor's application for an extension of time, can he reduce the period to which the contractor is entitled to reflect time saved by work omitted?

3.9.1   Architects and engineers when issuing variations which omit work often consider that where the variation shows a saving in time they are entitled to reduce the contract period or reissue an extension of time already granted but showing a shorter period.

3.9.2   Some contracts deal with this question. For example JCT 98 states in clause 25.3.6:

> 'No decision of the Architect under clause 25.3.2 or clause 25.3.3.2 shall fix a Completion Date earlier than the Date for Completion stated in the Appendix.'

It is clear that the architect cannot reduce the contract period irrespective of how many omissions he issues.

Where, however, an architect has granted an extension of time and is considering a further extension he can take into account omissions issued since the granting of the first extension of time. Clause 25.3.1 of JCT 98 dealing with this point states:

> 'the Architect shall, in fixing a new Completion Date, state: . . . the extent, if any, to which he has had regard to any instructions issued under clause 13.2 requiring as a Variation the omission of any work or obligation'

3.9.3   The ICE 6th and 7th Editions in clause 44(5) state that the engineer cannot decrease an extension of time already granted.

> 'No such final review of the circumstances shall result in a decrease in any extension of time already granted by the Engineer pursuant to sub-clauses (3) or (4) of this Clause'

There is no reference to taking into account time saved by omissions when granting extensions of time but this may be implicit in 'all the circumstances known to him' which the engineer considers in assessing the delay under 44(2)(a).

3.9.4   GC/Works/1 1998 states under condition 36(4):

> 'The PM shall not in a final decision to withdraw or reduce any interim extension of time already awarded, except to take account of any authorised omission from the Works or any relevant Section that he has not already allowed for in an interim decision.'

In this case the project manager cannot reduce the contract period but can subsequently reduce an extension of time already granted to take account of omissions.

3.9.5   GC/Works 1/Edition 2 states under condition 28:

> 'In determining what extension of time the contractor is entitled to the Authority shall be entitled to take into account the effect of any authorised omissions from the Works.'

3.9.6   MF/1, IChemE and the Engineering and Construction Contract (NEC) make no reference to taking into account omissions when extensions of time are considered. It would seem reasonable however for the engineer or supervising officer to take into account omissions when making a decision concerning an extension of time. The contractor however, having been given a period within which to carry out the work, e.g. the contract period or an extended contract, should not see the period subsequently reduced in the absence of express provision in the contract.

SUMMARY

Some forms of contract deal specifically with the question. In the absence of specific wording in the contract it is unlikely that a court would accept that an architect or engineer has power to reduce the contract period or any extension of time already granted. Nevertheless the contractor should continue to proceed diligently despite there being surplus time due to omissions since such periods can be taken into account if the need arises for a further extension of time.

3.10 Where a contractor's progress is behind programme, will he be entitled to an extension of time where the progress and completion is affected by exceptionally adverse weather but would not have been so affected if work had been on programme?

3.10.1 Many standard forms of contract, for example JCT 1998 clause 25.4.2, ICE 7th Edition clause 44(1)(d) and FIDIC 1999 Edition clause 8.4(c), include provisions which allow the contractor an extension of time for completion of the work where affected by exceptionally adverse weather or climatic conditions. In the case of *Walter Lawrence and Son Ltd* v. *Commercial Union Properties (UK) Ltd* (1984), the progress and completion of the work was delayed due to exceptionally adverse weather conditions. The progress of the work was behind programme and it was argued by the architect that, had work been carried out in accordance with the programme, the weather would not have affected the progress or completion of the works.

3.10.2 The court did not accept the architect's interpretation of the contract. It is clear from the wording of the contract, which was JCT 63, that if the delaying event, which in this case was exceptionally adverse weather, affects the progress and as a consequence the completion date, then a right to an extension of time arises. There is no reference in the extension of time clause in the contract to the programme and therefore the contractor's entitlement should not be affected if he is behind programme.

The wording of JCT 1998, ICE 7th Edition and FIDIC 1999 Edition is similar to that of JCT 63 and if applied to the circumstances which prevailed in this case would produce a similar result.

SUMMARY

If a contractor's progress is behind programme, it will not affect an entitlement to an extension of time where completion is delayed due to exceptionally adverse weather conditions. The governing factor is whether the exceptionally adverse weather delayed the actual progress and in turn the completion date.

4
Global Claims

4.1 Will a claim for an extension of time and the recovery of loss and expense which does not precisely detail the period of delay and the amount claimed in respect of each claim matter causing delay (i.e. a failure to link cause and effect) fail?

4.1.1 The proper manner of presenting a claim before a court or arbitrator is to link the cause of delay and extra cost with the effect. For example, if the architect or engineer is six weeks late in issuing the drawings for the foundations (cause) the completion date for completion of the work may, as a consequence, be delayed by six weeks (effect).

In recent time contractors and subcontractors have been ever willing to short cut the need to link cause and effect by use of the global claim. All causes of delay under the global claim method are lumped together and one overall delay given as a consequence. The usual requirement to link each cause of delay with its separately identified additional cost is ignored.

The preparation of claims on a global basis has been the subject of considerable controversy for a number of years. Cost savings can arise due to not having to rertain copious records, analyse them meticulously and spend time on research. It is, however, a difficult and developing area of law, the subject of a great deal of court scrutiny.

4.1.2 In support of the global claim, contractors and subcontractors draw comfort from the dicta in a number of legal cases.

In *J. Crosby & Sons Ltd* v. *Portland Urban & District Council* (1967), the contract overran by 46 weeks. The arbitrator held that the contractor was entitled to compensation in respect of 31 weeks of the overall delay, and he awarded the contractor a lump sum by way of compensation rather than giving individual periods of delay against the nine delaying matters. By way of justification the arbitrator in his findings said:

> The result, in terms of delay and disorganisation, of each of the matters referred to above was a continuing one. As each matter occurred its consequences were added to the cumulative consequences of the matters which had preceded it. The delay and disorganisation which ultimately resulted was cumulative and attributable to the combined effect of all these matters. It is therefore impracticable, if not impossible, to assess the additional expense caused by delay and disorganisation due to any one of these matters in isolation from the other matters.

The respondent contested that the arbitrator was wrong in providing a lump sum delay of 31 weeks without giving individual amounts in respect of each head of claim. Mr Justice Donaldson however agreed with the arbitrator saying:

> I can see no reason why he (the arbitrator) should not recognise the realities of the situation and make individual awards in respect of those parts of individual items of claim which can be dealt with in isolation and a supplementary award in respect of the remainder of these claims as a composite whole.

A similar award occurred in *London Borough of Merton* v. *Stanley Hugh Leach* (1985), where Mr Justice Vinelott said:

> The loss or expense attributable to each head of claim cannot in reality be separated.

4.1.3 This type of claim is now referred to as a 'global' or 'rolled up' claim. These decisions were thrown into question by *Wharf Properties Ltd and Another* v. *Eric Cumine Associates and Others* (1991). In this case the plaintiff made no attempt to link the cause with the effect in respect of a claim by the employer against his architect for failure properly to manage, control, co-ordinate, supervise and administer the work of the contractors and subcontractors as a result of which the project was delayed.

Six specific periods of delay were involved but the statement of claim did not show how they were caused by the defendant's breaches. The plaintiff pleaded that due to the complexity of the project, the interrelationship and very large number of delaying and disruptive factors and their inevitable knock-on effects, etc. it was impossible at the pleadings stage to identify and isolate individual delays in the manner the defendant required and that this would not be known until the trial.

The defendant succeeded in an application to strike out the statement of claim. The Court of Appeal in Hong Kong decided that the pleadings were hopelessly embarrassing as they stood (some seven years after the action began) and an unparticularised pleading in such a form should not be allowed to stand.

The matter was nevertheless referred to the Privy Council in view of the apparent differing view taken by the courts in *Crosby* and *London Borough of Merton*. The Privy Council, however, rejected the assertion that these two decisions justified an unparticularised pleading. Lord Oliver said:

> Those cases establish no more than this, that in cases where the full extent of extra cost incurred through delay depend upon a complex interaction between the consequence of various events, so that it may be difficult to make an accurate apportionment of the total extra costs, it may be proper for an arbitrator to make individual financial awards in respect of claims which can conveniently be dealt with in isolation and a supplementary award in respect of the financial consequences of the remainder as a composite whole. This has, however, no bearing upon the obligations of a plaintiff to plead his case with such particularity as is sufficient to alert the opposite party to the case which is going to be made against him at the trial. [The defendants] are concerned at this stage not so much with quantification of the financial consequences – the point with which the two cases

referred to were concerned – but with the specification of the factual consequences of the breaches pleaded in terms of periods of delay. The failure even to attempt to specify any discernible nexus between the wrong alleged and the consequent delay provides, to use [counsel's] phrase 'no agenda' for the trial.

4.1.4   The editors of *Building Law Reports*, Volume 52 at page 6 say, by way of observation:

> 'It must therefore follow from the decision of the Privy Council in *Wharf Properties* v. *Eric Cumine Associates* that *Crosby* and *Merton* are to be confined to matters of quantum and then only where it is impossible and impracticable to trace the loss back to the event. The two cases are not authority for the proposition that a claimant can avoid providing a proper factual description of the consequences of the various events upon which reliance is placed before attempting to quantify what those consequences were to him. Thus, taking the example before the Privy Council, it seems that it will in future be necessary for a plaintiff to be quite specific as to the delay which it is alleged was caused by an event such as a breach of contract or an instruction giving rise to a variation. This in turn will mean that those responsible for the preparation and presentation of claims of this kind will need to work hard with those who have first-hand knowledge of the events so as to provide an adequate description of them. Equally, it will mean that proper records will need to be kept or good use will have to be made of existing records to provide the necessary detail. It will no longer be possible to call in an outsider who will simply list all the possible causes of complaint and then by use of a series of chosen "weasel" words try to avoid having to give details of the consequences of those events before proceeding to show how great the hole was in the pocket of the claimant. There must be, as the Privy Council points out an "agenda" for the trial: there must be a discernible connection between the wrong and, where delays are relied on, the consequent delay.'

The Scott Schedule was originally designed to set out relevant points concerning defects in a manner which would easily be digested. It can however be adapted to suit any particular form of dispute where a great many disputed facts are involved.

4.1.5   In the case of *Imperial Chemical Industries* v. *Bovis Construction Ltd and Others* (1992), Judge Fox-Andrews QC ordered the plaintiff to serve a Scott Schedule containing:

- the alleged complaint
- the defendant against whom the claim was made
- which clause in the contract had been breached
- alleged failure consequences of such breach.

The Scott Schedule is not, however, a formula that can be applied to every case.

4.1.6   The whole subject came under review by the Court of Appeal in *GMTC Tools & Equipment Ltd* v. *Yuasa Warwick Machinery Ltd* (1994).

This case, relating to a defective computer controlled precision lathe (to be used in the manufacture of blanks which in turn were machined to become rotary cutters), had nothing to do with construction work but nonetheless the principles on which the decision was made will apply.

The lathe did not operate as intended and the plaintiff prepared and submitted a claim based on the number of management hours involved in dealing with the problem and the number of hours during which the lathe was inoperable.

Difficulties arose when the defendants sought further and better particulars of the claim. The judge ordered that a Scott Schedule should be drawn up providing detailed information attempting to link the cause (the malfunctioning of the lathe which caused down time) with its effect (the wasted management hours and the purchase of blanks to replace lost production).

Following attempts to reamend the Scott Schedule the matter came before the Court of Appeal after the plaintiff's failure to comply with an Unless Order.

The Court of Appeal was sympathetic to the plaintiff's situation. Lord Justice Leggatt said the defendant's argument presupposed that the plaintiff's production process was so flexible and instantaneously reactive to a period of down time that it would be possible to link each incident of down time with the purchase of a precise number of blanks to replace lost production. The opinions of the defendant were not accepted.

It was the view of Lord Justice Leggatt that a judge is not entitled to prescribe the way in which the quantum of damage is pleaded and proved. No judge, he said, is entitled to require a party to establish causation and loss by a particular method. Lord Justice Leggatt by way of conclusion said

> I have come to the clear conclusion that the plaintiff should be permitted to formulate their claims for damages as they wish, and not be forced into a strait-jacket of the judge's or their opponent's choosing.

4.1.7 A further review of global claims occurred in *British Airways Pension Trustees Ltd* v. *Sir Robert McAlpine & Sons* (1994) where a dispute arose out of the development of a site in Croydon.

There were defects in the work alleged to be due to faults by the architects, the contractor and others involved in the construction of the project and it was argued that the diminution in value of the property due to the defects was £3.1m which formed the basis of the claim, plus the cost of investigating the defects.

The defendants requested that further and better particulars be provided in respect of the claim. They asked to be given detailed information as to how much of the diminution in the value of the property could be attributed to each and every defect. For example, if two windows were defective how did it affect the price paid for the property?

They justified this type of question with the claim that until such details are given the defendant does not know the case to be answered and so faces an unfair hearing. On behalf of the plaintiff it was argued that all the defects had been identified and therefore due to their existence the project was worth £3.1m less than it would have been without the default.

An application was made by the defendants to strike out the claim and dismiss the action as they had been seriously prejudiced through a failure on the part of the plaintiffs properly to particularise their claim.

Judge Fox-Andrews ordered that the claim be struck out and the action dismissed. However, this decision was overruled by the Court of Appeal, Lord Justice Saville in summing up said:

> The basic purpose of pleadings is to enable the opposing party to know what case is being made in sufficient detail to enable that party properly to answer it. To my mind, it seems that, in recent years, there has been a tendency to forget this basic purpose and to seek particularisation even when it is not really required. This is not only costly in itself, but is calculated to lead to delay and to interlocutory battles in which the parties and the courts pore over endless pages of pleadings to see whether or not some particular points have or have not been raised or answered, when in truth each party knows perfectly well what case is made by the other and is able properly to prepare to deal with it. Pleadings are not a game to be played at the expense of citizens nor an end in themselves, but a means to the end, and that end is to give each party a fair hearing

This attitude is a precursor of the sweeping Woolf reforms which from April 1999 rewrote the court procedural rules to give judges a much more proactive role in the management of cases and cut down the interlocutory delaying tactics of the type described by Lord Justice Saville.

4.1.8   From the decision in *Amec Building Ltd* v. *Cadmus Investment Co Ltd* (1996) it seems that courts in the future will judge each case on its merits without laying down principles as to whether global claims will or will not be accepted.

In this case the judge's remarks as to the arbitrator's approach to global claims seem relevant:

> Certainly, it seems to me that there is no substance in the complaint that the arbitrator had set his face against global claims and that, thereby, prejudiced Amec. What appears to have happened, is that, upon justifiable complaint of lack of particularity, the arbitrator insisted upon an allocation of the overall claim to particular heads which was attempted by Amec and, when these matters were investigated by the accountants and in evidence and cross-examination, it clearly became quite clear to the arbitrator that there were occasions of duplications, matters compensated elsewhere and a general lack of particularisation. In those circumstances, it seems to be what the arbitrator concluded that the plaintiff had not proved the costs incurred were due to the fault by Cadmus. . . . As is clear from the careful judgment of the arbitrator, he proceeded to analyse each of the claims made by Amec and decided each upon the evidence that was before him.

4.1.9   In *Inserco Ltd* v. *Honeywell Control Systems* (1996) the court made an award based upon a global claim. The judge's comments make interesting reading:

> Inserco's pleaded case provided sufficient agenda for the trial and the issues are about quantification. Both *Crosby* [*Crosby* v. *Portland Urban District Council* (1977)] and *Merton* [*London Borough of Merton* v. *Stanley Hugh Leach* (1985)] concerned the application of contractual clauses. However, I see no reason in principle why I should not follow the same approach in the assessment of the amounts to which Inserco may be entitled. There is here, as in *Crosby* an extremely complex inter-action between the consequences of the various breaches, variations and additional works and, in my judgement, it is 'impossible to make an accurate apportionment of the total extra cost between the several causative events'. I do not think that even an artificial apportionment could be made – it would certainly be

extremely contrived – even in relation to the few occasions where figures could be put on time etc. . . . It is not possible to disentangle the various elements of Inserco's claims from each other. In my view, the cases show that it is legitimate to make a global award of a sum of money in the circumstances of this somewhat unusual case which will encompass the total costs recoverable under the February agreement, the effect of the various breaches which would be recoverable as damages or which entitle Inserco to have their total cost assessed to take account of such circumstances, and the reasonable value of the additional works similarly so assessed.

4.1.10 In *How Engineering Services Ltd* v. *Linder Ceilings and Partitions plc* (1999) the arbitrator, Mr Jupp, awarded a sum in respect of loss and expense based upon a global assessment. The dispute arose out of two contracts, the Atrium and the Station.

In finding for the claimant the arbitrator accepted the claimant's costs as set out in the points of claim and arrived at a figure of £130 346. From this he deducted £4186 in respect of work carried out prior to the receipt of the notice, £32 611 in respect of realignment of ceilings which was treated as a variation order, and £3155 for remedial work. The award was thus based on a total cost claim calculated on a global basis against which the defendant appealed on the basis that the arbitrator had not ascertained the sum as required by the arbitration clause.

It was the view of the court that, in some cases, the facts are not always clear. Different tribunals would reach different conclusions and an arbitrator is entitled to assess loss and expense in the same way as a court assessing damages. The court upheld the arbitrator's award.

4.1.11 In the case of *Berhards Rugby Landscapes Ltd* v. *Stockley Park Consortium* (1997), the plaintiff landscape contractor entered into an agreement under seal with Trust Securities Holdings (TSH) for the construction of a golf course on a landfill site under a contract which incorporated the ICE 5th Edition conditions. The work was subject to delay, and detailed and lengthy claims were submitted. It was alleged by the defendant that the claims were bound to fail due to a number of reasons, one of which was that they contained global delay claims for variations.

It was held by the court that the global claim was a total cost claim. The plaintiff had quantified its alleged loss by subtracting the expected cost of the works from the final costs. Such a claim was permissible if it was impractical to disentangle that part of the loss attributable to each head of claim and the situation had not been caused by the plaintiff's conduct. In such circumstances, the inference was that the employer's breaches had led to additional costs, and that the cause nexus was to be inferred rather than demonstrated.

4.1.12 The case of *John Doyle Construction Ltd* v. *Laing Management (Scotland) Ltd* (2004) was heard in the Scottish Inner House, Court of Session, and dealt with whether in principle global claims were bound to fail.

Laing Management was the management contractor for the construction of a new corporate headquarters for Scottish Widows in Edinburgh. A number of works packages were contracted out to John Doyle. An action was brought by John Doyle seeking an extension of time of 22 weeks together with loss and expense. John Doyle admitted that, despite their best efforts, it was impossible to identify cause and effect in respect of each item which caused delay and disruption and that was why the claim had been prepared on a global basis.

It was argued on behalf of Laing that the basis of the global claim was that all the additional costs incurred by John Doyle resulted from the delay and disruption caused by Laing. This being the case, if one of the events relied upon by Doyle was shown not to have been caused by Laing, then the case would be undermined. The judge was sympathetic to this view but felt that the whole matter should be treated with common sense. His view was that, if a global claim is to succeed, the contractor must eliminate from the claim all costs that are not the responsibility of the employer. He considered global claims to be a risky business but nonetheless allowed proceedings to continue.

4.1.13 The Society of Construction Law Delay and Disruption Protocol 2002 provides a view concerning global claims which accords with many of the judgments referred to:

> '1.14.1 The not uncommon practice of contractors making composite or global claims without substantiating cause and effect is discouraged by the Protocol and rarely accepted by the courts.
>
> 1.14.2 If the Contractor has made and maintained accurate and complete records, the Contractor should be able to establish the causal link between the Employer Risk Event and the resultant loss and/or expense suffered, without the need to make a global claim.
>
> 1.14.3 In what should only be rare cases where the financial consequences of the various causes of compensation are impossible to distinguish, so that an accurate apportionment of the compensation claimed cannot be made between the several causative events, then in this rare situation it is acceptable to quantify individually those items of the claim which can be dealt with in isolation and claim compensation for the remainder as a composite whole.
>
> 1.14.4 The Contractor will nevertheless need to set out the details of the Employer Risk events relied on and the compensation claimed with sufficient particularity so that the employer knows the case that is being made against it.'

SUMMARY

The complexity of contemporary claims sometimes needs to be dealt with by a 'global' approach, but this is not a carte blanche for the plaintiffs to put in any figure. Detail needs to be provided where it is available and can be asked for and/or ordered in a specific form, e.g. a Scott Schedule. However demands for particulars are not to be used as a delaying tactic and as an end in themselves.

Lord Justice Leggatt's views (in *GTMC Tools*) are contrary to the 'hands-on' case management style for judges at the core of the Woolf reforms, other strands of which are that initial claims should be as fully particularised as possible and interlocutory skirmishing (e.g. demands for further and better particulars and service of interrogatories) is to be kept to a minimum.

Will a claim which fails to link cause and effect fail? The answer must be that it depends upon the circumstances. It is unlikely however that the claim will be rejected entirely by the courts on the grounds that it has been prepared on a global basis; however any award is likely to be modest.

5

Liquidated and Ascertained Damages

5.1 What is the difference between liquidated damages and a penalty?

5.1.1 The terms liquidated damages and penalty are often interpreted wrongly. Liquidated damages are enforceable whereas a penalty is not.

Lord Dunedin had this to say of liquidated damages in the case of *Dunlop Pneumatic Tyre Co Ltd* v. *New Garage & Motor Co Ltd* (1915):

> The essence of a penalty is payment of money stipulated as *in terrorem* of the offending party; the essence of liquidated damages is a genuine covenanted pre-estimate of damage.

Lord Dunedin also went on to say:

> If the sum is 'extravagant and unconscionable in amount in comparison with the greatest loss that could conceivably be proved to have followed the breach' it will be regarded as a penalty and unenforceable.

5.1.2 In the case of *Public Works Commissioner* v. *Hills* (1966), in deciding what constituted a penalty, the judge said:

> The question whether a sum stipulated is a penalty or liquidated damages is a question of construction to be decided upon the terms and inherent circumstances of each particular contract, judged of as at the time of the making of the contract, not as at the time of the breach.

Where the stipulated liquidated damages are held to be a penalty, the employer will be able to recover only the amount of unliquidated damages he can prove.

5.1.3 It does not affect the situation as to whether the sum included in the contract is referred to as a penalty or liquidated and ascertained damages.

Keating on Building Contracts 5th edition at page 225 states:

> 'Though the parties to a contract who use the words "penalty" and "liquidated damages" may prima facie be supposed to mean what they say, yet the expression used is not conclusive. The court must find out whether the payment stipulated is in truth a penalty or liquidated damages.'

5.1.4 In the case of *Jeancharm Ltd* v. *Barnet Football Club Ltd* (2003) Jeancharm undertook to supply football kits to Barnet. If Barnet paid late they were required under the terms of the contract to pay interest at a rate of 5% per week. Jeancharm were obliged to pay 20p per garment for each day's delay in delivery. Claims were levied by both parties and the judge in the lower court ordered Barnet to pay Jeancharm the sum of £5000. It was argued by Barnet before the Court of Appeal that 5% per week equated to 260% per annum and was a penalty. It was argued by Jeancharm that in deciding whether 5% per week was a penalty the court should examine fully the risk borne by both parties which would include the 20p per garment per day for late delivery. The Court of Appeal held that on any basis 260% per annum interest should be regarded as a penalty.

5.1.5 Lord Justice Peter Gibson in the *Jeancharm* case considered there were four principles relating to liquidated damages which differentiated them from a penalty:

(1) The court had to look at the substance of the matter rather than the form of words used in order to identify the parties' intentions.

(2) The essence of a penalty would be distinct from a genuine pre-estimate of loss. In other words, the provision would not be construed as a penalty where a genuine pre-estimate had been carried out.

(3) The question of whether such a clause was to be treated as a penalty was a question of construction or interpretation of the contract at the time of making the contract, not at the time of the breach.

(4) If the amount was extravagant or unconscionable in comparison with the greatest loss that could be perceived, the clause would be a penalty.

SUMMARY

Liquidated and ascertained damages are a reasonable pre-estimate of the losses the employer is likely to incur if work is completed late. Such a sum is enforceable if due to his own default the contractor completes work late. A penalty on the other hand is a sum included in the contract which is intended to penalise the contractor and is far greater than the employer's estimated loss. Such a sum would be unenforceable.

5.2 If the employer suffers no loss as a result of a contractor's delay to completion, is he still entitled to deduct liquidated damages?

5.2.1 Contractors often argue that if they can show that when delays to completion occurred the employer suffered no loss or a substantially reduced loss then the liquidated damages expressed in the contract will not become payable.

5.2.2 The essence of liquidated damages is that they are a genuine covenanted pre-estimate of loss: *Clydebank Engineering and Shipbuilding Co* v. *Don José Yzquierdo & Castaneda* (1905).

It was said by Lord Woolf in the Hong Kong case of *Philips Hong Kong Ltd v. The Attorney General of Hong Kong* (1993):

> Since it is to [the parties'] advantage that they should be able to know with a reasonable degree of certainty the extent of their liability and the risk which they run as a result of entering into the contract. This is particularly true in the case of building and engineering contracts. In the case of those contracts provision for liquidated damages should enable the employer to know the extent to which he is protected in the event of the contractor failing to perform his obligations.

Liquidated damages are therefore a reasonable pre-estimate of the loss the employer anticipates he will suffer if the contractor completes late. Its advantage is that the contractors know in advance the extent of risks they are taking and employers do not have the expense and difficulty of proving their loss item by item.

5.2.3   In the case of *BFI Group of Companies Ltd* v. *DCB Integration Systems Ltd* (1987) a contract had been let using the JCT Minor Works Form to alter and refurbish offices and workshops. A dispute arose concerning liquidated damages and was referred to arbitration. The arbitrator held that there had been a delay in completion but declined to award liquidated damages on the grounds that the employer had suffered no resulting loss.

An appeal was lodged against the arbitrator's award and heard by Judge John Davies QC. He decided that the liquidated damages clause automatically came into play when the contractor completed late without a contractual justification and the employer was not required to demonstrate that he had suffered loss. The arbitrator was wrong in law in refusing to award payment of liquidated damages.

5.2.4   In the case of *Bovis Construction* v. *Whatling* (1995) it was held that a clause, such as a liquidated damages clause which limits liability, should state clearly and unambiguously the scope of the limitation and should also be construed with a degree of strictness.

SUMMARY

It was made clear by the decision in *BFI Group of Companies* v. *DCB Integration Systems Ltd* (1987) than an employer may, where they are provided for in the contract, deduct liquidated damages even though in the event he has suffered no loss.

5.3   If a delay is caused by the employer for which there is no specific entitlement to an extension of time expressed in the extension of time clause, will this result in the employer losing his right to levy liquidated damages?

5.3.1   One purpose for including an extension of time clause in a contract between employer and contractor is to provide a mechanism for adjusting the completion date where delays which affect completion are caused by the architect, engineer or employer and so preserve the employer's right to deduct liquidated damages in the event of further delay through the fault of the contractor.

However some contracts, including the JCT and ICE 6th Edition, do not provide within the extension of time clause for every conceivable type of delay which may be caused by the employer.

5.3.2   Vincent Powell-Smith, in *The Malaysian Standard Form of Building Contract* as at p88, had this to say on the matter:

> 'Clause 23 is gravely defective in many important respects and is in need of urgent amendment. The grounds on which an extension may be granted are very limited and do not cover many common delaying events e.g. failure by the Employer to supply materials to the contractor, failure to give agreed access and failure to give possession of the site on the due date. If such events occur and cause delay to completion the Architect has no power to grant an extension with the result that time will be "at large" and the Employer will lose his right to liquidated damages.'

The Malaysian Standard Form in this respect is identical to JCT 63.

In *Rapid Building Group* v. *Ealing Family Housing* (1984) the employer granted possession late and was prevented from levying liquidated damages in respect of delays subsequently caused by the contractor. The contract used was JCT 63. A similar situation arose in *Thamesa Designs SDN BHD* v. *Kuching Hotels SDN BHD* (1993).

JCT 98 clause 23.1.2 now overcomes one of the criticisms of Vincent Powell-Smith by providing for delay in the granting of possession for a period of up to six weeks.

5.3.3   Lord Justice Phillimore in the case of *Peak Construction (Liverpool) Ltd* v. *McKinney Foundations Ltd* (1970) said:

> I would re-state the position because I think it needs to be stated quite simply. As I understand it, a clause providing for liquidated damages [clause 22] is closely linked with a clause which provides for an extension of time [clause 23]. The reason for that is that when the parties agree that if there is delay the contractor is to be liable, they envisage that the delay shall be the fault of the contractor and, of course, the agreement is designed to save the employer from having to prove the actual damage which he has suffered. It follows, once the clause is understood in that way, that if part of the delay is due to the fault of the employer , then the clause becomes unworkable if only because there is no fixed date from which to calculate that for which the contractor is responsible and for which he must pay liquidated damages. However, the problem can be cured if allowance can be made for that part of the delay caused by the actions of the employer, and it is for this purpose that recourse is had to the clause dealing with extension of time. If there is a clause which provides for extension of the contractor's time in the circumstances which happen, and if the appropriate extension is certified by the architect, then the delay due to the fault of the contractor is disentangled from that due to the fault of the employer and a date is fixed from which the liquidated damages can be calculated.

5.3.4   To ensure that all delays by employer, architect or engineer are properly catered for in the extension of time clause fully comprehensive wording is required similar to GC Works/1 1998 which states under condition 36(2):

'The PM shall award an extension of time under paragraph (1) only if he is satisfied that the delay, or likely delay, is or will be due to –

. . .

(b) the act, neglect or default of the employer, the PM or any other person for whom the Employer is responsible . . .'

5.3.5 The ICE 7th Edition shows a change to the 6th Edition where in clause 44(1)(e) there is a comprehensive clause which covers all delays caused by the employer, the wording being 'any delay impediment prevention or default by the Employer'.

SUMMARY

If the contract does not provide grounds for extending the completion date due to Employer's delays employers who cause delay to completion will lose their rights to deduct liquidated damages in respect of the contractor's delays. Effective contracts avoid this by having a fully comprehensive extension of time clause.

5.4   Are liquidated damages based on a percentage of the contract sum enforceable?

5.4.1 It is not uncommon for contracts to include liquidated damages which have been calculated in accordance with some sort of formula. MF/1 provides in clause 34.1 for liquidated damages to be expressed as a percentage of the contract value. A limit to the contractor's liability is achieved by making provision for the amount of damages to be capped.

5.4.2 It has been argued that as the definition of liquidated damages is a genuine covenanted pre-estimate of loss (*Dunlop Pneumatic Tyre Co Ltd* v. *New Garage & Motor Co Ltd* (1915) damages based upon a formula which includes an estimate only of the value of the work cannot be a genuine pre-estimate of damage.

However Lord Dunedin in the *Dunlop* case made the following important comment:

> It is no obstacle to the sum stipulated being a genuine pre-estimate of damages that the consequences of the breach are such as to make precise pre-estimation almost an impossibility. On the contrary, that is just the situation when it is probable that pre-estimated damage was the true bargain between the parties.

5.4.3 Certain projects, for example a new road, a school or church, would present the parties to a contract with a difficult task in pre-estimating the loss for late completion. These are the types of projects which Lord Dunedin probably had in mind.

In *Robophone Facilities* v. *Blank* (1966) Lord Justice Diplock said:

> And the more difficult it is likely [to be] to prove and assess the loss which a party will suffer in the event of a breach, the greater the advantages to both parties of fixing by the terms of the contract itself an easily ascertainable sum to be paid in that event.

Lord Woolf in the same case said:

> The court has to be careful not to set too stringent a standard and bear in mind what the parties have agreed should normally be upheld. Any other approach will lead to undesirable uncertainty especially in commercial contracts.

5.4.4   In the case of *J.F. Finnegan* v. *Community Housing* (1993) the calculation of liquidated damages included the following formula

$$\frac{\text{Liquidated}}{\text{Damages}} = \frac{\text{Estimated Total Scheme Cost} \times \text{Housing Corporation Lending Rate} \times 85\%}{52}$$

The court held that the formula was a genuine attempt to estimate in advance the loss the defendant would suffer from late completion.

Judge Carr concluded by saying:

> I find that the formula used was justified at the time the parties entered into the contract.

SUMMARY

It would seem that provided:

- the calculation of an accurate estimate of liquidated damages is difficult or impossible
- the intention is not to penalise the contractor for completing late but to compensate the employer
- the formula is a genuine attempt to estimate in advance the employer's loss

a court will not refuse to enforce liquidated damages which have been calculated by applying a percentage to the contract sum.

5.5   **Where liquidated damages are expressed as so much per week or part thereof, and the contractor overruns by part of a week only but is charged a full week's liquidated damages, are the courts likely to consider this a penalty and therefore unenforceable?**

5.5.1   It has been suggested that liquidated damages expressed at a rate per week or part of a week cannot be a genuine pre-estimate of anticipated loss. The reasoning runs that the loss must differ for each additional day and therefore the same figure cannot apply for a day and also a full week.

5.5.2   Forecasting losses which may or may not occur in the future is not a precise science. Even where they are relatively easy to forecast they do not always increase proportionately with the passage of every day. For example an overrun to a completion date for a new office block may involve an employer remaining in existing premises for a longer period. It would be difficult to imagine a situation where extending an existing lease could be arranged on a daily basis. In all probability extensions to the lease for a three month or six month period is more likely to be the norm. Extending periods of employment are also unlikely to be arranged on a daily basis – a week or month would in all probability be the minimum period.

5.5.3   In *Philips Hong Kong Ltd* v. *The Attorney General of Hong Kong* (1993) a daily rate was included in the contract in respect of liquidated damages. The contract provided for the amount to be reduced if parts of the work were handed over in the delay period. A minimum daily rate was stipulated which would apply irrespective of how much of the work was handed over short of completion of all the work. It was argued that the minimum daily rate constituted a penalty and was thus unenforceable as a situation could be envisaged where the minimum daily rate well exceeded the employer's estimated losses. The court rejected the argument and upheld the sum included in the contract as liquidated damages.

SUMMARY

There is no report of liquidated damages expressed at a rate per week or part thereof ever having been argued before a court as being a penalty. It seems unlikely however that liquidated damages expressed in this form would be held to be a penalty and unenforceable merely on account of the manner in which the damages were expressed.

5.6   If the architect or engineer fails to grant an extension of time within a timescale laid down in the contract, will this prevent the employer from levying liquidated damages?

5.6.1   Many modern contracts such as JCT 98 and ICE 6th and 7th Editions lay down timescales within which extension of time awards are to be decided. (see 3.8.1).

- In JCT 98, by clause 25.3.1, the architect, if it is reasonably practicable having regard to the sufficiency of information submitted by the contractor, must make a decision within 12 weeks of the receipt of that information.
- The ICE 6th Edition requires the engineer under clause 44(5) to make decisions concerning extensions of time within 14 days of the issue of the certificate of substantial completion. The 7th Edition gives the engineer 28 days.
- GC/Works/1 1998 in condition 36(1) provides for the project manager to make a decision within 42 days of receiving the contractor's written notice.

Will a failure by the architect or engineer to comply with these timescales be fatal to the employer's right to deduct liquidated damages?

5.6.2   There have been two legal cases where this question has been considered. *Temloc* v. *Errill Properties Ltd* (1987) and *Aoki Corp* v. *Lippoland (Singapore) Pte Ltd* (1994).

5.6.3   The case of *Temloc* v. *Errill* arose out of a contract let using JCT 80. By the terms of this contract the architect is required to make decisions concerning extensions of time within a set timescale. With regard to the effect on the employer's entitlements should the architect fail to give his decision within the timescale, Lord Justice Croom-Johnson in the Court of Appeal had this to say:

> In my view, even if the provision of clause 25.3.3 [requirement for the architect to review extensions of time within 12 weeks of practical completion] is applicable, it is directory only as to time and is not something which would invalidate the calculation and payment of liquidated damages. The whole right of recovery of liquidated damages under clause 24 does not depend on whether the architect, over whom the contractor has no control, has given his certificate by the stipulated day.

5.6.4   A similar matter was the subject of the decision in *Aoki Corp* v. *Lippoland (Singapore) Pte Ltd* (1994).

Clause 23.2 of the SIA Conditions of Contract makes it a condition precedent that the contractor notifies the architect of any event, direction or instruction which the contractor considers entitles him to an extension of time. The architect is then required to respond in writing within one month indicating whether or not in principle the contractor is entitled to an extension of time. As soon as possible after the delay has ceased to operate and it is possible to decide the length of the extension, the architect will notify the contractor of his award. If the contractor fails to complete the work by the completion date or extended completion date, the architect must issue a delay certificate as soon as the latest date for completion has passed.

The contractor notified the architect of delays but the architect failed to notify the contractor of whether in principle an entitlement to an extension of time existed. Eventually, the architect, without giving his decision in principle, refused all requests for extension except one for which he allowed 15 days.

The employer deducted liquidated damages.

It was held:

- A decision by the architect on the principle of the contractor's right to an extension was not a condition precedent to a valid determination of the contractor's entitlement. The contractor could, however, claim damages as a result of the architect's failure to make a decision which might include the cost of increasing the labour force.
- There is no rule that delay in the issue of the delay certificate after the date for completion or the latest extended date for completion, renders the delay certificate invalid.

5.6.5   It would seem that failure by the architect or engineer to make a decision concerning extensions of time within a timescale laid down in the contract is not fatal to the employer's rights to deduct liquidated damages.

SUMMARY

Unfortunately contracts such as JCT 98 which provide a timescale within which the architect must grant an extension of time do not state what effect a failure to comply with the timescale will have upon the employer's rights to deduct liquidated and ascertained damages. However the decisions in *Temloc* v. *Errill Properties Ltd* and *Aoki Corp* v. *Lippoland* suggest that, provided a proper decision is made by the architect at some stage concerning extensions of time, a failure to meet the deadline will not affect the employer's rights.

5.7 If the contractor delays completion but no effective non-completion certificate is issued by the architect/engineer under a JCT contract, will this mean that the employer loses his right to deduct liquidated damages?

5.7.1 JCT 98 is an example of a contract which makes reference under clause 24.2.1 to the architect issuing a certificate when the contractor fails to complete on time. The question frequently asked is whether in the absence of the architect's certificate the employer remains entitled to deduct liquidated damages where the contractor finishes late.

5.7.2 This procedure was the subject of a decision of the High Court in the case of *A Bell & Son (Paddington) Ltd* v. *CBF Residential Care & Housing Association* (1989). A Bell, the contractors, entered into a contract with CBF Residential Care for the construction of an extension to a nursing home. The contract was JCT 80 Private Edition with Quantities with a date for completion of 28 February 1986. Liquidated damages for late completion were stated to be £700 per week.

5.7.3 Work commenced on time but completion was not achieved by 28 February 1986. The contractor served a delay notice and the architect granted an extension of time to provide a new completion date of 25 March 1986. Completion was not, however, achieved by this new date. At this stage the employer, CBF Residential Care, considered that as completion was late, liquidated damages would be due. JCT 80 requires the architect to issue a certificate of non-completion and provides for the employer to write to the contractor indicating an intention to deduct liquidated damages. Both architect and employer complied with this procedure. However the architect subsequently had second thoughts and granted two further extensions of time. The first extended the completion date to 14 April 1986, the second further extended the date to 21 April 1986. Unfortunately the contractor did not complete the work until 18 July 1986 when the architect issued a certificate of practical completion. A long delay then occurred before the architect on 3 December 1987 granted another extension of time extending the completion date to 20 May 1986. There was still, however, a shortfall between the date of 20 May 1986 by which time the contractor should have completed at 18 July 1986 when practical completion was achieved.

5.7.4   The architect issued a final certificate on 25 February 1988 but the balance due was reduced by £4 900 before payment in respect of liquidated damages. It was argued by the contractor that liquidated damages should not have been deducted as the procedures required by JCT 80 had not been properly complied with. Following his first granting of an extension of time showing a revised date for completion of 14 April 1986 the architect had issued a non-completion certificate indicating that the contractor had failed to achieve this date but following the grant of further extensions of time the non-completion certificate was not re-issued to reflect the revised dates for completion. It was argued that it should have been and, in the absence of properly re-issued non-completion certificates, the employer lost the right to deduct liquidated damages. The architect had issued a final certificate and it was therefore too late to re-issue the non-completion certificate.

In finding in favour of the contractor and ordering that the £4 900 be paid, plus interest and costs, the court held:

> Construing clause 24.1 strictly and in accordance with its plain and ordinary meaning, it demands the issue of a certificate when a contractor had not completed by 'the completion date' ... I think that when a new completion date is fixed, if the contractor has not completed by it, a certificate to that effect must be issued, and it is irrelevant whether a certificate has been issued in relation to an earlier, now superseded completion date ...
>
> Construing clause 24.2.1 in a similar manner to clause 24.1, since the giving of a notice is made subject to the issue of a certificate of non-completion, if the certificate is superseded, then logically the notice should fall with it ... If a new completion date is fixed, any notice given by the employer before it is at an end.

Accordingly the condition precedent to the permissible deduction of liquidated damages, i.e. the issue of an architect's non-completion certificate, had not been fulfilled and the employer therefore lost the right to deduct liquidated damages.

5.7.5   The matter of a non-completion certificate was again referred to in *J.F. Finnegan* v. *Community Housing* (1993) when it was held that a written notice from the employer under JCT 80 is a condition precedent to the right to deduct liquidated damages.

5.7.6   The wording has been tightened up in JCT 98 where in clause 24.2.1 it states:

> 'Provided:
> – the Architect has issued a certificate under clause 24.1
> . . .
> The Employer may ...
> require in writing the Contractor to pay to the Employer liquidated and ascertained damages ...'

SUMMARY

The employer will lose the right to deduct liquidated damages where JCT 98 applies if the architect fails to issue a proper non-completion certificate under clause 24.1.

5.8 Can a subcontractor who finishes late have passed down to him liquidated damages fixed under the main contract which are completely out of proportion to the subcontract value?

5.8.1   This question presumes that the subcontractor has a contractual obligation to finish within a timescale and is in breach of the obligation if he completes late. Where a subcontractor is in breach he will have a liability to pay damages to the main contractor.

The general principles covering damages for breach of contract are explained in *Hadley* v. *Baxendale* (1854) and later fully considered in *Victoria Laundry (Windsor) Ltd* v. *Newman Industries* (1949).

5.8.2   Briefly the injured party is entitled to recover any loss likely to arise in the usual course of things from the breach, plus such other loss as was in the contemplation of the parties at the time the contract was made and which is likely to result from the breach.

The contractor, as injured party, is entitled to levy a claim for damages against a subcontractor who completes late. These damages should include only those losses which under normal circumstances are likely to arise and are within the contemplation of both parties. In all probability a court would hold that the contractor's claim should include his own additional costs plus any legitimate claims received from the employer and other subcontractors who have suffered financially as a result of the sub-contractor's late completion. If the normal standard forms of contract are employed the employer will levy a claim for liquidated damages against the main contractor if the main contract completion is delayed due to a default on the part of a subcontractor. Under normal circumstances these liquidated damages will form a part of the main contractor's claim against the defaulting subcontractor irrespective of the value of the subcontract works.

5.8.3   This rule will apply in all cases except where the subcontractor is nominated and the terms of the main contract provide the main contractor with an entitlement to an extension of time where delays are caused by a nominated subcontractor's default. Delays by the nominated subcontractor would result in an extension of time being granted to the main contractor and hence no claim from the employer for liquidated damages.

5.8.4   Where the sum for liquidated damages under the main contract could be classed as out of the ordinary and therefore not within the contemplation of the subcontractor, it may be argued that the subcontractor is obliged to reimburse the main contractor only that element of the employer's liquidated damages which is normal and usual. Two problems arise out of this type of argument. Firstly, the question of what we mean as normal and usual; and secondly, if the sum for liquidated damages is so out of the ordinary, it might be regarded as a penalty and unenforceable.

5.8.5   Usually main contractors will send to subcontractors, with the tender enquiry, details of the main contract including the sum for liquidated damages. This procedure prevents subcontractors from arguing that the sum was outside their contemplation when they entered into the sub-contract.

5.8.6   One way out for subcontractors is to include in the subcontract an amount
for liquidated damages which provides a cap on their liabilities.

In *M.J. Gleeson plc* v. *Taylor Woodrow Construction Ltd* (1989) Taylor
Woodrow were management contractors for work at the Imperial War
Museum and entered into a subcontract with Gleeson. The management
contract provided for liquidated damages at £400 per day and clause 32 of
the subcontract provided for liquidated damages at the same rate. Clause 11
(2) of the subcontract also provided that if the subcontractor failed to
complete on time the subcontractor should pay:

> 'a sum equivalent to any direct loss or damage or expense suffered or incurred by
> [the management contractor] and caused by the failure of the subcontractor. Such
> loss or damage shall be deemed for the purpose of this condition to include for any
> loss or damage suffered or incurred by the authority for which the management
> contractor is or may be liable under the management contract or any loss or
> damage suffered or incurred by any other subcontractor for which the manage-
> ment contractor is or may be liable under the relevant subcontract.'

Gleeson finished late and they received from Taylor Woodrow a letter as
follows:

> 'We formally give you notice of our intention under clause 41 to recover monies
> due to ourselves caused by your failure to complete the works on time and dis-
> ruption caused to the following subcontractors. The following sums of money are
> calculated in accordance with clause 11(2) for actual costs we have incurred or may
> be liable under the management contract.'

Then followed a summary of account showing deductions of £36 400 for
liquidated damages, being £400 per day from 31 May 1987 to 31 August
1987, and £95 360 in respect of 'set-off' claims from ten other subcontractors.

Gleeson applied for summary judgment under Order 14 in respect of the
retained sum of £95 360 and were successful. Judge Davies found that Taylor
Woodrow had no defence:

> On the evidence before me, therefore, Taylor Woodrow's course of action against
> Gleeson in respect of set-offs is for delay in completion. It follows that it is included
> in the set-off for liquidated damages, and to allow it to stand would result in what
> can be metaphorically described as a double deduction.

SUMMARY

Subcontractors who, in breach of their subcontract, complete late will be
liable to pay the resultant damages incurred by the contractor. These
damages will include any liability the main contractor has to pay liquidated
damages to the employer which result from the delay. This procedure will
apply irrespective of the value of the subcontract works.

It is open to the subcontractor to argue, if the main contract liquidated
damages are extremely high, that the sum involved was outside his
contemplation at the time the contract was entered into. To forestall this type
of argument main contractors, usually with the tender enquiry documents,

will set out details of the main contract (including the sum included for liquidated damages).

Where the subcontractor is nominated and the main contract provides for an extension of time where work is delayed by the subcontractor no claim from the employer for liquidated damages will arise provided that the contractor has properly claimed the extension of time.

5.9   What is meant by 'time at large'? How does it affect the employer's entitlement to levy liquidated damages for late completion?

5.9.1   'Time at large' means there is no time fixed for completion or the time set for completion no longer applies.

Agreements for work to be carried out are often entered into without a completion period being stated. Letters of intent often contain instructions to commence work without a completion date being agreed. In these cases time is said to be 'at large'.

5.9.2   Contractors can find themselves trapped into contracts where the time allowed for completion is too short and the amount of money to which they are entitled is insufficient to meet their additional costs. In these circumstances they may turn to alternative means of rectifying the situation other than the normal claims for extensions of time and additional payment.

5.9.3   For some time contractors have used the 'time at large' argument in an attempt to avoid paying liquidated damages. Their normal approach is to say that the contract period has either never been established or that, due to delays caused by the employer for which there is no express provision in the contract for extending the completion date, time becomes at large (see 5.3). This being the case, the contractor's obligation is merely to finish within a reasonable time.

5.9.4   The contractor successfully used this argument in the case of *Peak Construction (Liverpool) Ltd* v. *McKinney Foundations Ltd* heard before the Court of Appeal in 1970. It was held that, as delays on the part of the City Council in approving remedial works to the piling were not catered for in the extension of time provisions, the right to liquidated and ascertained damages was lost and time became at large. The Corporation was left with an entitlement to claim such common law damages as a result of the contractor failing to complete within a reasonable time as it was able to prove.

5.9.5   The case of *Rapid Building Group* v. *Ealing Family Housing*, heard before the Court of Appeal in 1984, involved a contract let using JCT 63. Unfortunately, due to the presence of squatters, the housing association was unable to give possession of the site to the contractor on the due date. There was no provision in JCT 63 for extensions of time for late possession. The contractor was therefore able to argue successfully that time became at large. The obligation was altered to completing within a reasonable time and the employer lost its rights to levy liquidated and ascertained damages.

5.9.6   In the case of *Inserco Ltd* v. *Honeywell Control Systems* (1996) Inserco contracted to complete all work by 1 April 1991. Due to additional and revised work, and lack of proper access and information, Inserco was prevented from completing on time. There was no provision in the contract for extending the completion date and time was held to be at large.

5.9.7 If time does become 'at large', the contractor's obligation is to complete within a reasonable time. What is a reasonable time is a question of fact: *Fisher* v. *Ford* (1840). Calculating a reasonable time is not an easy matter and would depend on the circumstances of each case. As *Emden's Building Contracts'* 8th edition puts it in Volume 1 at page 177:

> 'Where a reasonable time for completion becomes substituted for a time specified in the contract ... then in order to ascertain what is a reasonable time, the whole circumstances must be taken into consideration and not merely those existing at the time of the making of the contract.'

5.9.8 Vincent Powell-Smith in his book *Problems In Construction Claims* at page 78 has this to say concerning 'time at large':

> If for some reason time under a building contract becomes 'at large', the Employer can give the contractor reasonable notice to complete within a fixed reasonable time, thus making time of the essence again: *Taylor* v. *Brown* (1839). However, if the contractor does not complete by the new date, the Employer's right to liquidated damages does not revive, and he would be left to pursue his remedy of general damages at common law.'

5.9.9 Time may also become at large where the architect or engineer fails properly to administer the extension of time clause as required by the contract. An example would be where an architect or engineer fails to make any award where a proper entitlement exists.

SUMMARY

Time is at large when a contract is entered into with no period of time fixed for completion. Where this occurs the contractor's obligation is to complete work within a reasonable time.

There may also be circumstances which arise rendering a completion period fixed by the contract as no longer operable, again rendering time at large. An example is where a delay is caused by the employer and the terms of the contract make no provision for extending the completion date due to delays by the employer.

5.10 Can a contractor challenge the liquidated damages figure included in a contract as being a penalty and unenforceable after the contract is signed? If so, will it be a matter for the employer to prove the figure to be a reasonable pre-estimate of anticipated loss?

5.10.1 A golden rule when interpreting contracts is that both parties are bound by the terms of the contract into which they enter. With this in mind can a contractor, having signed the contract which includes a sum for liquidated damages, later, challenge the figure as being a penalty?

Two fairly recent cases have brought the question of liquidated and ascertained damages and their enforceability into focus: *Philips Hong Kong* v. *The Attorney General of Hong Kong* (1993) and *J. Finnegan Ltd Community Housing Association* (1993).

5.10.2   Both contracts included a clause relating to liquidated and ascertained damages. In the *Finnegan* case liquidated damages were stated to be £2 500 per week or part thereof. The *Philips* case involved a contract in which the liquidated damages varied between HK$ 77 818 per day and HK$ 60 655 per day depending on how much of the works had been handed over.

The parties in both cases entered into the contracts without the contractors challenging the sums included as liquidated damages at that time.

5.10.3   In the *Finnegan* case a housing association let a contract based upon JCT 80 to construct eighteen flats. The contractual date for completion was 1 March 1988 and the liquidated and ascertained damages were fixed at £2 500 per week or part thereof.

The contractor failed to complete the works by the contractual date for completion and liquidated and ascertained damages of £47 500 for the period 1st March to 13th August were deducted from monies due to the contractor.

The contractor then challenged the liquidated damages as being a penalty and unenforceable. The court held that the figure (although based on a formula) was a genuine pre-estimate of loss and therefore enforceable (see 5.4.4). It was, however, at no time suggested that the contractor was unable to challenge the liquidated damages amount on the ground of his having signed the contract.

5.10.4   In the *Philips* case the contractor again made a late challenge to the liquidated damages figure on the grounds that it was a penalty. Again there was no difficulty in leaving it until the end of the day. However the contractor, in like manner to Finnegan, was unable to demonstrate that the liquidated damages clause was a penalty and so unenforceable.

In arriving at a decision in the *Philips* case the court was influenced by *Robophone Facilities Ltd* v. *Blank* (1966) where Lord Justice Diplock stated that:

> The onus of showing that a stipulation is a penalty clause lies upon the party who is sued upon it

In other words the contractor facing a claim for liquidated damages which he challenges as being a penalty is put to proof that his allegation is correct. It is not for the employer to prove that the liquidated damages figure is a reasonable pre-estimate of loss.

SUMMARY

A contractor who enters into a contract which contains a liquidated damages figure can at a later stage challenge the amount as being a penalty and unenforceable. However where he makes such a challenge it is up to him to demonstrate that the amount is a penalty and not a reasonable pre-estimate of the employer's loss. It is not for the employer to justify the figure.

5.11 If liquidated damages to be enforceable must be a reasonable pre-estimate of loss, how can public bodies or organisations financed out of the public purse be capable of suffering loss?

5.11.1 It is an established principle that for liquidated damages to be enforceable the sum claimed should be a genuine pre-estimate of anticipated loss. From this, contractors often argue that organisations such as health authorities, education trusts and the like are financed from the public purse and as such can never suffer loss. This argument if successful would be unsavoury to any right-minded person.

5.11.2 The effect of public money was at issue in a construction case *Design 5* v. *Keniston Housing Association Ltd* (1986). In this case the plaintiffs were a firm of architects and the defendant a registered housing association. The plaintiffs sued for unpaid fees. It was argued by way of defence that failures on the part of the plaintiffs in their design, supervision and contract administration had resulted in an increase in expenditure amounting to £14m. In answer to the counterclaim the architects maintained that, whether or not they had been at fault and whether or not the costs of the scheme had been increased, the housing association had suffered no loss. This was because the housing association was entitled to receive Housing Association Grant (HAG) from the Department of the Environment in a sum equal to the actual cost of the scheme regardless of any fault by the architects.

Judge Smout, in finding for the housing association, said:

> It is pertinent to note that the general rule, that only nominal damages can be awarded where there has been a wrong but no loss, has never been absolute. Various exceptions are as old as the rule itself, others have developed piecemeal...
>
> In this respect it is sufficient to echo the comments expressed in the argument of the defendants, namely that the purpose of Housing Association Grants is to provide housing for the needy, and not to be used to relieve professional advisers from the financial consequence of breach of contract and negligence.

The housing association did not get a 'windfall' as the grant was reduced accordingly.

SUMMARY

It would seem that it is not open to a contractor or professional adviser who is liable for breach of contract to argue that, as the employer is publicly funded and hence incurs no loss, only nominal damages should be awarded.

5.12 If liquidated damages become unenforceable and hence an entitlement to unliquidated damages arises, can the unliquidated damages be greater than the liquidated damages?

5.12.1 An extension of time clause serves two purposes. In the first instance it enables a contractor to be relieved from the obligation to complete on time if events occur which would otherwise be at his risk – for example excessively adverse weather provided for in JCT 98. Secondly an extension of time clause provides a mechanism for adjusting the completion date to take account of delays caused by the architect, engineer, employer and those employed or engaged by the employer. In the event of there being no such provision, time can become at large and the right to levy liquidated damages for any delays caused by the contractor is lost if delays are to any of these reasons.

This is not a satisfactory situation. In *Rapid Building* v. *Ealing Family Housing* (1994) Lord Justice Lloyd commented

> Like Lord Justice Philimore in *Peak Construction (Liverpool)* v. *McKinney Foundations Ltd* (1969) I was somewhat startled to be told in the course of the argument that if any part of the delay was caused by the employer, no matter how slight, then the liquidated damages clause in the contract, clause 22 becomes inoperative.

In *Peak Construction* it was held:

> If the Employer is in any way responsible for the failure to achieve the completion date, he can recover no liquidated damages at all and is left to prove such general damages as he may have suffered.

5.12.2 Where liquidated damages become unenforceable it will be for the employer to prove such general or unliquidated damages as he is claiming. This is in contrast to liquidated damages which require no proof to be enforceable. The question then arises as to whether, should the general damages which the claimant is able to prove exceed the liquidated damages included in the contract will payment become due for the greater amount so proved.

5.12.3 There is little in the way of case law concerning this matter. In the old case *Wall* v. *Rederiaktiebolaget Luggude* (1915) it was held that, where a liquidated damages figure was held to be inappropriate, the unliquidated damages which were proved to have been incurred could be levied in full even though they exceeded the amount of liquidated damages. In contrast a more recent Canadian case, *Elsley* v. *Collins Insurance Agency Ltd* (1978), provided an opposite view. The question of whether a general damages claim is limited to the amount of liquidated damages included in the contract was referred to in *Cellulose Acetate Silk Co Ltd* v. *Widnes Foundry* (1933) but was left open.

5.12.4 As the case law on this subject is sparse and inconsistent, it may be relevant to apply an old rule to the situation, namely that a party cannot benefit from his own breach.

In *Alghussein Establishment* v. *Eaton College* (1988) it was held:

> It has been said that, as a matter of construction, unless the contract clearly provides to the contrary it will be presumed that it was not the intention of the

parties that either should be entitled to rely on his own breach of duty to avoid the contract or bring it to an end or to obtain a benefit under it.

It would seem that in the light of this decision an employer who caused a delay for which there was no provision for an extension of time and so rendered time at large should not be able to recover in respect of the contractor's delays general damages which exceed the liquidated damages stated in the contract.

SUMMARY

There is little consistant authority on this point but is seems unlikely that unliquidated damages would be held to be enforceable to an extent which exceeds the amount of liquidated damages.

5.13 Where a contract includes a single liquidated damages amount for failing to complete the whole of the works by the completion date, what entitlement does the employer have to claim from the contractor who has failed to complete parts of the work by the milestone dates written into the contract?

5.13.1 Most standard forms of contract provide for a date by which all work should be completed. If the contractor fails to complete the work by that date, then he is in breach of contract. JCT 1998, ICE 7th Edition, FIDIC 1999 Edition and GC/Works/1 (1999), together with most other standard forms of contract, provide the employer with a right to deduct a liquidated damages sum written into the contract for each day or week the contractor is late completing. An exception to the norm can be found in the ECC. Where this contract is used, liquidated damages for delay are not provided for unless Option J – Delay Damages applies. Where no provision for liquidated damages is included in the contract, the contractor, if he completes late, will be liable for unliquidated damages. For these to apply, the employer will be required to produce proof of the losses he has sustained as a result of the contractor's failure to complete on time.

5.13.2 The employer may, in addition to including a date by which all of the work must be completed, provide for the handover of the work in stages. Handover dates are usually provided and some standard forms of contract include special conditions which apply. JCT 1998, for example, has a Sectional Completion Supplement, whilst ICE 7th Edition includes in the Form of Tender provision for completion dates for sections of the work, sometimes referred to as milestone dates. With this type of arrangement separate liquidated damages amounts may be inserted into the contract in respect of each section. A failure on the part of the contractors to finish work by the section completion date will leave him open to a claim from the employer for the recovery of liquidated damages.

5.13.3 Where the contract provides for section completion dates but there are no liquidated damages stated, then if the contractor completes a section or more than one section late, he will be liable to pay the employer unliquidated damages.

5.13.4   It is sometimes the case that there is a liquidated damages sum written into the contract for a failure on the part of the contractor to complete the whole of the works, but no liquidated damages included for failing to achieve the sectional completion dates. Under these circumstances, if the contractor is late in completing the whole of the works, he will be obliged to pay to the employer liquidated damages. Failure to complete by the sectional completion dates will leave him exposed to a claim from the employer to pay unliquidated damages.

In claiming liquidated damages for failure on the part of the contractor to complete by milestone dates, the employer may be challenged to demonstrate as to whether, at the time of tender, they represented a reasonable pre-estimate. In some cases, it may be impossible to show that very much loss could be expected to result from a failure to meet the milestone dates. This being the case the employer will be unable to apply the liquidated damages.

SUMMARY

Contractors often include a single liquidated damages figure in respect of the whole of the works but no liquidated damage figures for the milestone completion dates. The employer will be entitled to levy a claim for liquidated damages if the contractor fails to complete the whole of the works by the completion date, and for unliquidated damages for a failure to complete by the milestone dates. However, the employer will be required to prove the losses he has incurred due to the contractor's failure to complete by the milestone dates.

5.14   Is it possible to include in a subcontract an all embracing sum for liquidated and ascertained damages for delay to completion?

5.14.1   It is common practice for a daily or weekly sum to be included in a standard main form of contract in respect of liquidated damages. These damages become payable by the main contractor to the employer in the event of a delay to completion of the works. The sum for liquidated damages must be a reasonable pre-estimate of the loss the employer is expected to incur should the work be completed late.

5.14.2   It is rare, however, to find a standard form of subcontract which includes provision for the deduction of liquidated damages should the subcontractor complete late. The Public Sector Partnering Contract Option 8 is however an exception. The normal procedure is for the main contractor to levy a claim against the subcontractor in respect of actual loss incurred. The delay by the subcontractor may result in delay to the completion of the main contract works, resulting in the employer levying against the contractor liquidated damages. The contractor would then pass down the liquidated damages claim to the subcontractor. The contractor may also incur costs of his own resulting from the subcontractor's delay, which may include the cost of accommodation, supervisory staff and plant retained on site during the delay period. These costs will usually be added to the employer's liquidated claim damages and charged against the subcontractor.

5.14.3   One of the difficulties for the subcontractor is that when pricing for the work he has no idea what the financial liability will be in the event of his late completion. If the subcontract delay does not affect the main contract completion date, the main contractor's claim may be modest. Alternatively, a substantial claim can be expected if the subcontractor's delay affects the main contract completion date. This type of uncertainty can be avoided if a liquidated and ascertained damage amount is included in the subcontract. The contractor, however, may be in a dilemma in calculating the daily or weekly sum. Should it be assumed that any delay by the subcontractor will automatically affect the main contract completion date and therefore the amount when calculated ought to include the liquidated damages in the main contract? If this is the case, subcontractors may be put off from submitting a tender by the size of the resultant amount.

5.14.4   The main contractor may decide to include the liquidated damages from the main contract when calculating the amount to be included in the subcontract. This figure could then be included in several subcontracts as delays on the part of any of those subcontractors could affect the main contract completion date. If, in the final analysis, many weeks' delay to each of these subcontracts occurred, which did not affect the main contract completion date, the claims from the main contractor would result in a windfall profit. Would courts be prepared to enforce claims of this nature? The view may be taken that it would be unreasonable to assume that all subcontract delays would affect the main contract completion date. Therefore only those subcontracts in respect of work which was on the critical path current at the time the subcontract was entered into should provide for a subcontract liquidated damage amount which includes the liquidated damages in the main contract.

SUMMARY

There should be no impediment to the inclusion of a sum for liquidated and ascertained damage in subcontracts. The Public Sector Partnering Contract Option 8 Subcontract includes for such a provision. The main contractor may experience some difficulties with enforcement if, in calculating the liquidated damages, the sum includes the liquidated damages under the main contract. This may well be the case if subcontract work was not on the critical path current at the time the subcontract was entered into.

6
PROGRAMME

6.1 Where a contractor submits a programme which is approved or accepted by the architect/engineer, is he obliged to follow it or can he amend it at his own discretion?

6.1.1 The programme is usually intended to be a flexible document. If the contractor gets behind, say due to the insolvency of a subcontractor, he would normally expect to revise the programme in an attempt to make up lost time. For this reason programmes are rarely listed as contract documents. It is the requirement of most contracts that obligations provided for in contract documents must be carried out to the letter. With a programme containing some hundred or more activities, compliance with the start and finish date for each without the possibility of revision would be impractical. For this reason programmes should not be contract documents.

6.1.2 Nevertheless some forms of contract will not permit the contractor to amend its programme once accepted without approval. For example, GC/Works/1 1998 condition 33(2) states:

> 'the contractor may at any time submit for the PM s agreement proposals for the amendment of the Programme.'

MF/1 clause 14.4 is worded along similar lines to the effect that the engineer's consent is required before the contractor can make any material change to the programme.

6.1.3 Clause 14(4) of the ICE 6th and 7th Editions empowers the engineer to require the contractor to produce a revised programme if progress of the work does not conform with the accepted programme. The revised programme must show the modifications to the accepted programme to ensure completion on time. That apart there is no restriction placed upon the contractor who wishes to revise his accepted programme.

6.1.4 The Engineering and Construction Contract (formerly the NEC) clause 32.1 calls on the contractor to show

> 'on each revised programme
>
> - the actual progress achieved on each operation and its effect upon the timing of the remaining work
> - the effects of implemented compensation events and of notified early warning matters

- how the Contractor plans to deal with any delays and to correct notified Defects; and
- any other changes which the Contractor proposes to make to the Accepted programme.

Clause 32.2 goes on to add that the contractor is required to submit a revised programme to the project manager for acceptance.

6.1.5   Other forms of contract, for example JCT 98, do not expressly require the contractor to seek approval to the amendment of his programme. However if amendments are made without approval the architect may however feel under no obligation to issue drawings to meet the revised programme.

SUMMARY

Some forms of contract require the contractor to seek approval or acceptance before amending his programme, for example GC/Works/1 1998, MF/1 and the Engineering and Construction Contract.

In the absence of an express requirement to seek approval to amend, the contractor can revise his programme as he wishes. An architect or engineer who has not been asked to approve or accept an amended programme may feel under no obligation to issue drawings in good time to enable the contractor to comply with the revised programme.

6.2   Is a subcontractor obliged to follow a main contractor's programme?

6.2.1   Most standard forms of contract provide for the contractor to produce a programme. A failure on the part of the contractor to produce the programme amounts to a breach of contract. It is not usual, however, for a contract to state expressly that a contractor must follow the programme. An exception is GC/Works/1 1998 which states in condition 34(1) that the contractor

> 'shall ... proceed with diligence and in accordance with the Programme or as may be Instructed by the PM ...'

6.2.2   It is unusual for a programme to be classified in a contract as a contract document. If it were so then contractors would be required to carry out work strictly in accordance with the programme. This could prove very exacting and in many instances impossible.

6.2.3   The situation with subcontractors is similar to that of a main contractor. An example of the obligation of a subcontractor with regard to a main contractor's programme occurred in *Pigott Foundations* v. *Shepherd Construction* (1994).

Pigott was employed as a domestic subcontractor to design and construct bored piling on a new fourteen storey office block. The main contract was JCT 80, the subcontract DOM/1, and Shepherd Construction the main contractor.

Pigott's subcontract provided for work to be carried out in eight weeks. Piling work commenced on 26 June 1989. However the bulk of the work was not finished until 20 October 1989. Pigott then left site and returned in April 1990 to carry out the remaining nine piles.

After commencement, work had proceeded at a slow pace with only one pile completed in the first week and further difficulties arose due to piling work which was alleged to be defective. It was not clear whether this was due to faulty design or bad workmanship and Pigott claimed that the difficulties arose as a result of ground conditions. A solution to the problem was reached which involved installing additional piles.

It was decided in this case that where DOM/1 conditions apply a subcontractor is not required to comply with the main contractor's programme.

6.2.4   If there exists an obligation for a subcontractor to carry out work to suit a main contractor's programme it can be a two edged sword for the main contractor. Such a requirement would place an obligation upon the main contractor to provide access to enable the subcontractor to carry out the subcontract work in accordance with the main contractor's programme. Contractors often experience difficulties in this respect as happened in the case of *Kitson Sheet Metal Ltd* v. *Matthew Hall Mechanical & Electrical Engineers Ltd* (1989). The court had to decide whether Kitsons, the subcontractors, were entitled under the contract to work to the programme and whether any written order requiring departure from it constituted a variation.

It was held that the parties must have recognised the likelihood of delays and of different trades getting in each other's way and that the prospects of working to programme were small. Provided Matthew Hall had done their best to make areas available for work they were not in breach of contract even if Kitsons were brought to a complete stop. Kitsons were therefore unable to recover the additional cost due to a substantial overrun on the contractor's programme.

A similar situation occurred in the case of *Martin Grant & Co Ltd* v. *Sir Lindsay Parkinson & Co Ltd* (1984). Again the court held that there was no entitlement for the subcontractor to claim extra due to delays to the main contract programme.

SUMMARY

A subcontractor is not required to follow a main contractor's programme unless provided for expressly in the terms of the subcontract; equally the main contractor is not obliged to grant access etc. to enable the subcontractor to do so.

6.3   Who owns float time in the contractor's programme, the architect/engineer or the contractor?

6.3.1   Most prudent contractors will allow some form of contingency in their programme. Risk analysis is becoming a frontline skill in construction projects. More of the risk and hence uncertainty is being placed upon contractors. Bad ground, strikes, weather conditions, shortages of labour and materials are now regularly allocated in the contract as the contractor's risk. Contractors and their subcontractors often make mistakes which have to be corrected. The contractor therefore will be unwise not to make provision in his programme for these uncertainties.

A prudent contractor will always include an element of float in his programme to accommodate these variables.

6.3.2 The question however is this – if the contractor has clearly programmed an activity to take longer than is estimated to complete, can the employer take advantage of the float time free of cost? This might prove useful if the architect/engineer is late issuing drawings or delays have been caused by the employer himself.

It may be argued that float will not be on the critical path and so the employer using it will not cause any delay or disruption. Hence the contractor will not become entitled to compensation.

6.3.3 Nevertheless Keith Pickavance in his book *Delay and Disruption in Construction Contracts* at page 335 makes reference to a case heard before the Armed Services Board of Contract Appeals in the USA (*Heat Exchanges* (1963)). Here it was held that the contractor's original cushion of time (which was not necessary for performance) should still be preserved when granting an extension of time for employer caused design delays. In an earlier case the Army Corporation of Engineer's Board of Contract Appeals had recognised the contractor's right to reprogramme, thereby giving him the benefit of the float. American courts also took the line on a management dispute that

> Total float may be used to programme jobs for all contractors; free float belongs to one contractor for programming any one activity . . .

> Neither total float nor free float is to be used for changes. (*Natken and Co* v. *George A Fuller & Co* (1972).

6.3.4 It would seem that in this country it is unlikely for an arbitrator to award an extension of time if the employer's delay did not affect the completion date. However, most arbitrators would take a sympathetic view to a contractor who reprogrammes to overcome a delay in the early part of the contract due to his own errors or risk items and in so doing uses up the float in the latter part of the programme.

Float which the employer may wish to have taken advantage of has thus disappeared.

6.3.5 The matter of float time was in evidence in the case of *Ascon Contracting Ltd* v. *Alfred McAlpine Construction* (1999). McAlpine was the main contractor for the construction of a five-storey building known as Villiers Development in Douglas, Isle of Man, near to the sea front. Ascon was appointed as subcontractor for constructing the reinforced concrete floor slabs, basement perimeter walls and upright columns between floors. The subcontract period was 29 weeks commencing on 28 August 1996, with completion by 5 March 1997.

Practical completion of the subcontract work was not achieved until 9 May 1997, nine weeks late. No extension of time was granted. Ascon submitted a

claim for an extension of time and payment of £337 918. McAlpine counterclaimed £175 000 liquidated damages paid to the employer, plus its own
loss and expense.

McAlpine's claim against Ascon in respect of liquidated damages alleged
to have been paid to the employer in the sum of £175 000 left the judge
unimpressed. A dispute between McAlpine and the employer had been
compromised by a final payment of £9475 inclusive of all McAlpine's claims
and explicitly no deduction was made for liquidated damages.

Part of McAlpine's case against Ascon was that, had all subcontractors
started and finished on time and McAlpine executed their own work on
time, practical completion would have been achieved 5 weeks early.
McAlpine's argument was that the 5 weeks float was for their benefit to
absorb their own delays. As the 5 weeks had been used by Ascon and other
subcontractors, McAlpine claimed they were entitled to recover their lost
benefit. The judge rejected this argument. He considered the float to be of
value in the sense that delays could be accommodated in the float time. This
would avoid an overrun to the contract period and hence any liability to pay
liquidated damages to the employer. The judge went on to say that McAlpine, whilst accepting the benefit against the employer, could not claim
against the subcontractors.

The judge seems to be taking the view that if float time is available it can be
used on a first come first served basis.

6.3.6   The Society of Construction Law Delay and Disruption Protocol 2002
expresses an opinion concerning float time which is not dissimilar to the
judgment in the *Ascon* case, where it states:

> 'unless there is express provision to the contrary in the contract, where there is
> remaining float in the programme at the time of an Employer Risk Event, an EOT
> should only be granted to the extent that the Employer Delay is predicted to reduce
> to below zero the total float on the activity paths affected by the Employer Delay.'

What is being said is that if a contractor becomes entitled to an extension of
time due to an employer's delay, any remaining float time must be taken up
first and only if there is then an anticipated delay to completion will an
extension of time be relevant. Any contractor delays which occur after all
float has been used and which affect the completion date will create a liability on the part of the contractor to pay liquidated damages to the
employer.

SUMMARY

There is no hard and fast rule but it would seem that, as a contractor will
normally include float in his programme to accommodate his risk items
which cannot be accurately predetermined in terms of time involvement,
and also to provide time for correcting mistakes, then the float belongs to
him and the employer or architect/engineer cannot object if later repro-
gramming by the contractor absorbs it. There is however a conflicting view
expressed in the Society of Construction Law Delay and Disruption Protocol

2002 which states that float belongs to the project and may be used on a first come first served basis.

6.4 What is the effect of making the programme a contract document?

6.4.1 The standard forms in general use require the contractor to produce a programme.

JCT 98 clause 5.3.1.2 requires the contractor to produce two copies of his master programme as soon as possible after the execution of the contract. No details are given as to whether a bar chart will suffice or if a critical path network is required.

GC/Works/1 1998, condition 33(1) is more precisely worded:

> 'The Contractor warrants that the Programme shows the sequence in which the Contractor proposes to execute the Works, details of any temporary work, method of work, labour and plant proposed to be employed, and events, which, in his opinion, are critical to the satisfactory completion of the Works; that the Programme is achievable, conforms with the requirements of the Contract, permits effective monitoring of progress, and allows reasonable periods of time for the provision of information required from the Employer; and that the Programme is based on a period for the execution of the Works to the Date or Dates of Completion.'

ICE 6th and 7th Editions in clause 14 require the contractor to submit a programme to the Engineer for approval within 21 days of the award of the contract. As with JCT 98 no reference is made to the type of programme required.

6.4.2 Whilst these standard forms require a programme to be provided, the programme itself is not listed as one of the contract documents. This is a very sensible arrangement. If a programme were to be given the status of a contract document, the contractor would be required to comply with it to the letter. All flexibility which is the key to catching up when progress gets behind would disappear.

6.4.3 However sometimes, however, by accident or design, a programme or method statement becomes a contract document. This was the situation in *Yorkshire Water Authority v. Sir Alfred McAlpine & Son (Northern) Ltd* (1985).

'The plaintiff invited tenders for a tunnel at Grimwith Reservoir. The contract was to incorporate the ICE 5th Edition and clause 107 of the specification stated:

> '*Programme of Work:* In addition to the requirement of clause 14 of the conditions of contract, the contractor shall supply with his tender a programme in bar chart or critical path analysis form sufficiently detailed to show that he has taken note of the following requirements and that the estimated rates of progress for each section of the work are realistic in comparison with the labour and plant figures entered in the Schedule of Labour, Plant and Sub-Contractors...'

The defendant submitted a tender in the standard ICE form accompanied by a bar chart and method statement. The method statement was approved at a meeting and two month's later the defendant's tender was accepted by letter. A formal agreement was signed incorporating, *inter alia*, the tender, the minutes of the meeting, the approved method statement and the plaintiff's letter of acceptance.

The method statement had followed the tender documents in providing for the construction of the works upstream. The contractors maintained that in the event it was impossible to do so and, after a delay, the work proceeded downstream. The contractors contended that, in the circumstances, they were entitled to a variation order under clause 51(1) of the ICE conditions. The dispute was referred to arbitration.

The arbitrator made an interim award in favour of the contractor and the employer appealed.

It was held by the court:

(1)   The method statement was not the programme submitted under clause 14 of the contract.
(2)   The incorporation of the method statement into the contract imposed on the contractors an obligation to follow it in so far as it was legally or physically possible to do so.
(3)   The method statement therefore became the specified method of construction so that if the variation which took place was necessary for completion of the works, because of impossibility within clause 13(1), the contractors were entitled to a variation order under clause 51 and payment under clause 51(2) and 52.

6.4.4   A similar situation arose in *English Industrial Estates Corporation* v. *Kier Construction Ltd and Others* (1992).

The instructions to tenderers provided that the tender should be accompanied by a full and detailed programme indicating the tenderer's proposed work sequence together with a brief description of the arrangements and methods of demolition and construction which the tenderer proposed to adopt for the carrying out of the contract works. In due course, Kier prepared a method statement and enclosed it with their tender.

It provided that details for

> on site crushings of suitable demolition arisings, removal of unsuitable arisings and excavation material, and filling excavations with suitable crushed arisings.

Disputes arose as to whether or not the contractors, through their sub-contractors, were entitled to export off site materials excavated which they thought uneconomic to crush. This dispute, *inter alia*, was referred to an arbitrator, whose decision was upheld by the court. His award was that Kier's method statement was a contract document, but nevertheless the wording was such that under it the contractor remained free:

• to decide which arisings it was uneconomic to crush; and
• to import replacement fill in place of hard arisings he chose not to
• crush; and
• to export uncrushed hard arisings.

6.4.5 In *Havant Borough Council* v. *South Coast Shipping Company Ltd* (1996) a method statement was given the status of a contract document. The contractor was unable to follow the method statement due to a court injunction which restricted the hours of working. To overcome the problem which involved excessive noise the contractor worked a different system to that provided for in the method statement. The court held that this constituted a variation and he was entitled to be paid.

6.4.6 The Engineering and Construction Contract (NEC) includes original wording which is to be contrasted to wording traditionally used in construction contract. Clause 31.2 deals with the contractor's obligations to submit a programme 'within the period stated in the Contract Data' unless the programme has been 'identified in the Contract Data'.

It is not clear what is the status of 'Contract Data' and whether it would be considered by the courts as having the same effect as a contract document in the ICE 6th and 7th Editions.

The courts are likely to say that unless the contract clearly states that the contractor is duty bound to carry out work strictly in accordance with the programme then no such obligation exists.

6.4.7 Programmes provided by many contractors are drawn up with claims in mind. Often unrealistic timescales are given within which drawings and instructions for the expenditure of prime cost and provisional sums are to be produced. Architects and engineers need to be very watchful at the time programmes are submitted to ensure that proper notice is given to the contractor of their feelings as appropriate that the programme is unrealistic with regard to the timing of the architect/engineer's functions.

Unfortunately there are no sanctions provided in the contract where a contractor neglects to provide a programme as required. In the absence of express wording, the architect/engineer would be in breach in refusing to allow a contractor who had not produced a programme to start work. The exception is the Engineering and Construction Contract (NEC) which provides in clause 50.3 for only 25% of payments due to be made until the contractor fulfils his obligation to provide a programme.

SUMMARY

If a programme is given the status of a contract document the contractor will be obliged to follow it to the letter. Should events not due to the contractor's own negligence arise which make it impossible to follow the programme, then it is likely that a court would award the contractor a variation and additional cost necessary to overcome the problem.

7
PAYMENT

7.1 Where a contract requires the contractor to give a guaranteed maximum price does he have any grounds for increasing the price above the guaranteed maximum?

7.1.1 The term guaranteed maximum price when applied to a construction contract provides for the employer a nice feeling of security. When entering into a contract of this nature, the employer is convinced that no matter what happens the final cost will not be above the maximum and there is a fair chance it could be lower. Any design changes which result from the specific instructions of the employer would understandably fall outside the guaranteed price. Guaranteed maximum price contracts have been in use for many years. IDC, a Stratford on Avon construction company who pioneered design and construct contracts some 25 or 30 years ago, promoted their contracts as guaranteed maximum price. It is a good selling point which can be persuasive. With risk transfer being very much in vogue, the guaranteed maximum price contract has seen a resurgence. The intention of this procurement route is to transfer all risks to the contractor and allow for no increase in price whatsoever other than costs which result from employer changes.

Construction is unlike most other commercial processes such as the motor car industry where a prototype is built and when perfected mass produced on an assembly line. Each construction project is a one-off and unique. The designer has to get it right first time and the manufacture and on-site process allows for no serious errors. The contractor who undertakes a guaranteed maximum price contract has to take these risks, together with risks such as unforeseen ground conditions, unexpected encounter with service mains, bad weather, industrial unrest, shortages of labour, plant and materials, changes in legislation, insolvency of suppliers and subcontractors, fire, storm and earthquake. Contractors are even often required to check all information received from the employer and take responsibility for its accuracy. It is hardly surprising that when one or more of these risks becomes a reality and substantial additional costs are incurred, the contractor casts round to find good legal reasons for receiving payment above the maximum price.

7.1.2 The case of *Mowlem* v. *Newton Street Limited* (2003) illustrates the difficulties which can befall a contractor who enters into a guaranteed maximum price contract. Work involved the conversion of a post office, built of reinforced concrete in 1910 in Manchester, into 104 apartments and an underground car park, with commercial units at ground level. The contract was an amended standard form. Article 10 expressly stated that the parties agreed that the contract sum was a guaranteed maximum price and that the contractor acknowledged he had taken all risks and responsibilities. Under a heading of contractor's risk the contractor also became responsible for any incorrect or insufficient information given to him by any person, whether or not in the employment of the employer. This represented a high level of risk in view of the fact that the work involved the conversion of a building which was almost 100 years old.

Difficulties arose as a result of the issue by the employer's agent of an instruction for the contractor to carry out concrete repairs to the existing structure. It was argued on behalf of Mowlem that as there was no specific reference in the Employer's Requirements or the Contractor's Proposals to the concrete repairs, they were entitled to be paid for the work over and above the guaranteed price. The lawyers representing Newton Street Limited were of the opinion that Article 10, which placed all risks onto Mowlem, deprived them of any right to additional payment. It was the view of the judge in finding against Mowlem that within the scheme of the guaranteed maximum price there is nothing to displace the ordinary and unambiguous meaning of Article 10 that the risk of unforeseen defects in the existing building was the contractor's.

7.1.3   The question needs to be asked as to whether it was reasonable and sensible for a contract involving the conversion of an old post office to be let on a guaranteed maximum price basis. There was no knowing what the contractor might have encountered. The existing structures and foundations, despite a proper precontract survey, may have proved to be insufficiently robust to carry the load and to have required strengthening. Asbestos may have been discovered, which could have resulted in a substantial cost to remove it. Commercial organisations who purchase old property cannot sensibly expect to transfer all the risk of the suitability of the property for conversion on to the contractor. In attempting to price the risk in a sensible manner, the contractor is likely to end up taking a gamble. When an unpriced risk becomes a reality, the contractor, rather like Mowlem, will inevitably read the fine print to see if there is a get-out. Some types of contract more easily lend themselves to be let on a maximum price basis, where, for example, the contractor is able to properly assess at tender stage the full extent of the work necessary to meet the employer's requirements.

7.1.4   There are no standard forms of contract which are classified as guaranteed maximum price. The one which comes the nearest is the FIDIC Silver Book for use on international projects, and although not titled a guaranteed maximum price contract, this is its obvious intention. Employers on domestic projects who require a guaranteed maximum price are left with either amending an existing standard form or having a bespoke contract produced. Contractors who market themselves on the basis of guaranteed maximum price usually produce their own form of contract. In view of the problems encountered by Mowlem, it is hardly surprising that, despite the title, contractors will usually include somewhere in the fine print reasons for increasing the price. A contract which includes reference to a guaranteed maximum price therefore is not always what it appears.

SUMMARY

It is not uncommon for a contract to include in its heading 'Guaranteed Maximum Price'. The fact that this heading exists does not in itself mean that the contractor, except for employer changes, has taken on all risks. It is usually necessary to read carefully through the wording of the contract to establish that the employer has transferred all risks to the contractor. It is not unusual to find that matters such as increases in cost due to changes in legislation fall outside the guaranteed price and will be paid for as an extra.

7.2 Where a subcontract provides for $2\frac{1}{2}$% cash discount, does this mean that discount can only be deducted if payment is made on time or may the discount be taken even if payment is made late?

7.2.1 Most subcontractors when submitting quotations include for $2\frac{1}{2}$% discount. Standard forms of subcontract, such as DOM/1 and NSC/C used with JCT (98), provide for $2\frac{1}{2}$% discount but clearly state that the discount is in return for payment on time. Where non-standard terms apply what does the term 'cash discount' mean – discount for payment on time or a trade discount which in essence is a price reduction and applies whether or not payment is made on time?

7.2.2 Subcontractors frequently complain that many contractors pay late but still deduct cash discount. Contractors often argue that they are entitled to deduct cash discount irrespective of whether they have paid on time or not.
The dictionary definitions of cash discount make its meaning clear:

- *New Collins Concise Dictionary:* 'Cash discount. A discount granted to a purchaser who pays before a stipulated date.'
- *Encyclopaedia of Real Estate Terms* 1987 edition: 'Cash discount. A reduction in price or consideration for early or prompt payment.'

7.2.3 Despite the apparent clarity of these definitions disputes have nevertheless been referred to court concerning cash discounts. In one there was, however, an unexpected slant which affects the discount provisions in many of the standard subcontract forms.
The case was *Team Services Plc* v. *Kier Management & Design* (1992). Kier were management contractors for the construction of a shopping and leisure centre at Thurrock with Team Services appointed as a subcontractor.
In a letter to Team Services dated 23 March 1989 Kier wrote:

'We hereby accept your fixed price tender sum of £14 811 000 . . . all in accordance with the attached schedule marked Appendix A inclusive of 2.5% cash discount and exclusive of VAT.'

The payment terms within the subcontract dated 7 January 1991 read as follows:

> 'Within 7 days of the receipt of the payment due under the principal agreement pursuant to a certificate which showed a sum in respect of the subcontract works the management contractor ... discharge such sum less:
>
> (i)   retention money at the rate specified in Appendix A
> (ii)   a cash discount of $2\frac{1}{2}\%$ on the difference between the said total value and the said retention, and
> (iii)   the amount previously paid.'

7.2.4   It was the subcontractor's case that on a proper interpretation of the wording in the contract the management contractor was entitled to deduct $2\frac{1}{2}\%$ cash discount from the interim certificates only if payments had been made within the timescale provided for within the subcontract. Alternatively the subcontractor considered there was an implied term to that effect.

The management contractor's case was that, on a true construction of the terms of the subcontract, the discount could be deducted irrespective of when the interim payments were made. In any event it was argued that it would have been a simple matter to have included appropriate wording if the intention was for discount to have been deducted only when all payments were made on time. Further it was suggested that, as the retention could be deducted even if payment was made late, the same would apply to discount. Judge Bowsher disagreed. The word 'cash' he thought significant and it should be given due weight. Discounts he considered are given for a purpose, if not, it would be simpler merely to reduce the price.

In finding for the subcontractor it is interesting to note the following matters which Judge Bowsher took into account in interpreting the meaning of cash discount:

(1)   The object is to ascertain, by the contractual words used the mutual intentions of the parties as to the legal obligations each assumed: *Pioneer Shipping Ltd* v. *BTP Tioxide Ltd* (1982).
(2)   The parties cannot give evidence themselves as to their intentions when drafting the contract. What should be ascertained is what the court considers a reasonable person's intentions would have been if placed in the position of the parties who drafted the contract – an objective rather than subjective test: *Reardon-Smith Line Ltd* v. *Hansen Tangren* (1976).
(3)   The words of a contract should be construed in their ordinary and grammatical sense: *Grey* v. *Pearson* (1857).
(4)   Courts may consult dictionaries in order to determine the meaning of words: *Pepsi Cola* v. *Coca Cola* (1942).

7.2.5   The subject of 'cash discount' was again raised in *Wescol Structures Ltd* v. *Miller Construction Ltd* (1998) where it was held that discount could only be deducted if payments were made on time.

7.2.6   Discounts expressed as a cash discount or stipulated as being offered in return for payment in accordance with the contract terms cannot be deducted if payment is late. Trade discounts on the other hand may be deducted irrespective of when payment is made.

7.2.7   It is well worth noting that if discount can only be deducted if payment is made on time contractors may be tempted to calculate discounts by reference to the gross amount due before the deduction of previous payments. If this method is used the contractor need only make the final payment on time to become entitled to deduct the full discount from the gross value of work executed. The contractor would have to forego discounts from interim payments made late. If he decided to pay late but illegally deducted discount the contractor's liability would be to pay interest on the overdue money.

The methodology would involve:

(1)   Deduct discount off the gross payment before deducting previous payments.
(2)   Make all interim payments late and forego the discount.
(3)   Pay the final payment on time and deduct discount from the gross final account

e.g.   Final account   100,000
        Less 2½% discount   2,500
                                                        97,500
        Less previous payment   90,000
        Balance due   £   7,500

The net result will be that the main contractor will have paid all interim payments late but because the final payment was made on time discount can be deducted.

SUMMARY

From the decision in *Team Services* it is clear that the term 'cash discount' means a discount which can only be deducted if payment is made on time. This case is, on the face of it, good news for subcontractors in that late payment will result in the contractor losing his cash discount.

Unfortunately for subcontractors the court held that as discount, in accordance with the wording of the subcontract in question and, indeed, many of the standard forms of subcontract, is deducted from the gross value, the position can be rectified by timely payment of a subsequent valuation.

The advice to main contractors is that they should always make sure the final payment is on time and for subcontractors to vary the subcontract terms in this regard – if they are in a strong enough position to do so.

7.3   Under what circumstances are contractors/subcontractors entitled to be paid for materials stored off site as part of an interim payment or payment on account?

7.3.1   Most standard forms of contract provide for interim payments on a regular basis usually monthly or by way of stage or milestone payments. It is the norm to include within those payments sums to reflect the value of work executed and also materials brought onto or adjacent to the site but unfixed. Usually the materials are required to be adequately stored and protected.

7.3.2　Contractors and subcontractors often suffer from a heavy drain of money in respect of materials and components which are manufactured but stored off site due to their size or value or due to restricted storage space on site. Special provision in the contract is necessary before interim payment can be made for these items.

7.3.3　A problem from the point of view of the employer would result from payment for materials or components stored off site if the contractor or subcontractor became insolvent before they were delivered to site. Lengthy and expensive disputes could arise between employer and the receiver or liquidator both as to the identity of the goods and components and also as to ownership (see 7.15).

7.3.4　JCT 98 by clause 30.3 provides for materials goods or prefabricated items stored off site to be included in an interim payment at the discretion of the architect. On rare occasions only does the architect ever exercise this discretion. Before they are eligible for inclusion in an interim payment certificate the following conditions need to be fulfilled:

- The contractor must provide reasonable proof that he has ownership and that ownership will vest in the employer after payment.
- The contractor must provide a bond unless the items are uniquely listed and the Appendix does not require a bond in favour of the employer.
- The items in storage must be set apart and properly marked.
- The contractor must provide proof of insurance.

7.3.5　ICE 6th Edition, clause 65 (1)(c) and 7th Edition, clause 60(1)(c), provide for interim payment for materials or goods stored off site which are identified in an appendix to the form of tender.

7.3.6　GC/Works/1 1998, condition 48 C, allows for payment of 'Things Off Site', but only if stated in the abstract of particulars, and this is not applicable in Scotland. This condition requires proof of ownership and its transfer to the employer, identification, proper storage, etc. in like manner to JCT 98.

7.3.7　The Engineering and Construction Contract (NEC) makes no reference to payment of goods stored off site.

SUMMARY

Payment for materials stored off site will only occur if the contract makes specific provision for such payment and the requisite conditions are fulfilled.

7.4　Can a contractor force an employer to set aside retention money in a separate bank account?

7.4.1　Most standard forms of construction contract provide for the employer to withhold sums of money referred to as retention money until completion.

The contractor is at risk should the employer become insolvent prior to payment of the retention.

In an effort to protect contractors and nominated subcontractors clause 30.5.1 of JCT 98 stipulates that the employer's interest in such a retention is fiduciary, as trustee for the contractor and for any nominated subcontractor. Other JCT forms are similarly worded.

7.4.2   The protection given to contractors in respect of retention held in trust was tested in *Rayack Construction Ltd* v. *Lampeter Meat Co Ltd* (1979). The conditions of contract were JCT 63 with wording in clause 30(4) similar to JCT 98. It was the decision of the court that the employer was obliged to set aside the retention money in a separate trust fund.

JCT 98 under clause 30.5.3 reinforces the decision in the *Rayack Construction* case in requiring the employer to place retention money in a separate bank account if requested by the contractor or any nominated subcontractor.

7.4.3   The whole subject of withholding retention money under a JCT contract may seem simple but a number of cases have highlighted certain difficulties.

7.4.4   In *Wates Construction (London) Ltd* v. *Franthom Properties Ltd* (1991), a contract was let for the construction of a hotel at in Kent. The contract conditions were JCT 80 With Quantities and the contract sum £2,869,607. The Employer was entitled to deduct 5% as retention in the case of some works, the total amount retained representing $2\frac{1}{2}$% out of the total sum.

Work started in June 1988 and by March 1990, when the court proceedings had commenced, the amount certified was £3 368 912 and the employer had retained £84 222. The works were certified as practically complete by in August 1989.

In accordance with clause 30.5.3 of the conditions of contract, the employer should have deposited the retention monies in a separate bank account. On 2 October 1989, the contractor became aware that retention monies were not being placed in a separate bank account. The contractor requested the employer to comply but he refused. A writ was issued by the contractor claiming an injunction to compel the employer to place the retention monies in a separate account. Judge Newey, as Official Referee, decided that the contractor was entitled to the order. The employer considered the judge to be wrong and lodged an appeal.

The issues considered were as follows.

(1)   Under clause 30.1.1.2, the employer has a right of deduction from retention monies for such matters as liquidated damages. The employer therefore contended that he was a beneficiary of the trust and under no obligation to place retention monies in a separate bank account. This argument was dismissed by the Court of Appeal as they considered the wording of clause 30.1.1.2:

> 'Notwithstanding the fiduciary interest of the Employer in the Retention as stated ... the Employer is entitled to exercise any right under this Contract of ... deduction from monies due ...'

They held that this suggests that the interest of the employer in the retention remains as a trustee and is not that of beneficiary. The Court of Appeal considered that in respect of the total retention fund of £84 222 there were five beneficiaries, the main contractor and four nominated subcontractors.

(2) The employer further argued that he was only bound to appropriate a sum to the retention fund and was not bound to set it aside in the sense of placing it in a separate identifiable account. It was the employer's contention that he could use the money in his own business as working capital. Clause 30.5.2.1 requires the architect or, if he so instructs, the quantity surveyor, to prepare a statement specifying the contractor's retention and the retention for each nominated subcontractor. The preparation of the statement, it was argued, was a sufficient appropriation of the sums involved. It was held by the Court of Appeal that the first duty of a trustee is to safeguard the fund in the interests of the beneficiaries. It would therefore be a breach of trust for any trustee to use the trust fund in his own business.

(3) The most powerful argument put forward on behalf of the employer was that clause 30.5.3, which requires him to put the retention in a separate account, had been deleted from the contract. It was said that the deletion of the clause by the parties was indicative of a common intention on their part that the employer should be under no obligation to place retention monies in a separate bank account. The Court of Appeal considered that the parties may have had differing reasons for wishing clause 30.5.3 to be deleted and therefore it was not possible to draw from the deletion of that clause a settled intention of the parties common to each of them that the general duties of the employer were other than those set out in clause 30.5.3 i.e. fiduciary and that this could include placing the monies in a separate account if requested to do so.

The Court of Appeal dismissed the appeal from the employer and upheld Judge Newey's order that a sum of £84 222 be deposited in a separate bank account.

7.4.5 The judge in *Herbert Construction* v. *Atlantic Estates* (1993) took a different view. In this case reference in JCT 80 to retention money being placed in a separate bank account in like fashion to the *Wates* case was deleted. It was held, unlike the *Wates* decision, that the employer was not bound to place retention money in a separate account.

7.4.6 In *Finnegan Ltd* v. *Ford Sellar Morris Developments Ltd* (1991), a contract was let using JCT 81 With Contractor's Design. Relying on the decision in *Wates* v. *Franthom Properties* the plaintiff sought to compel the defendant to place retention monies in a separate account. The defendant failed to do this, arguing that the plaintiff's demand of 22 April 1991 (which was 15 months after practical completion) had been made too late to comply with the provisions of JCT 80 clause 30.5. This states that the request should be made either at the commencement of the contract or before the issue of the first interim certificate.

It was held that JCT 81 clause 30.5 places no restriction on the time in which a contractor should make a request for retention to be placed in a separate trust fund. Further, it was not sensible to expect the contractor to make a request each time retention is added to a monthly certificate.

7.4.7 In *MacJordan Construction* v. *Brookmount Erostin* (1991) the contractor acted too late to preserve the retention fund. The employer who was a property developer became insolvent and had not placed retention in a separate bank account. MacJordan's claim to be paid retention from sums available did not take precedence over a bank floating charge. The amount of retention which the contractor was unable to recover totalled £109 000.

7.4.8 Subcontractors under certain circumstances are also protected. *In PC Harrington Contractors Ltd* v. *Co-Partnership Developments Ltd* (1998), Lelliott contracted with the plaintiff as works contractor under the Works/Contract/2 subcontract under the JCT 87 Management Form.

By clause 4.1 of the conditions of contract , Co-Partnership Developments, the employer under the contract and the defendant in the action were obliged to pay to Lelliott the cost of the project, which included sums due to Harrington. Amounts paid by Co-Partnership Developments were subject to the withholding of retention in respect of works by both the management contractor and the works contractor.

Lelliott went into receivership shortly after practical completion and following the expiry of the defects liability period. As a result of the insolvency Lelliott's employment was automatically determined. Co-Partnership Developments Ltd, held the retention fund of £288 166, which was expressed as being held on trust for the management contractor and for any works contractor in clause 4.8.1 of the management contract.

The dispute arose as a result of Co-Partnership Developments setting off additional costs it incurred due to Lelliott's insolvency from retention money held on behalf of Harrington. The court directed that the sum held by the employer under clause 4.8.1 was held in trust on behalf of the works contractor. Co-Partnership Developments were not, therefore, entitled to set off costs incurred as a result of the insolvency of Lelliott from retention held on behalf of Harrington.

SUMMARY

Where the JCT conditions apply, employers can be forced by a court injunction to place the retention money in a separate bank account. It is not necessary for the contractor to make a request for retention to be paid into a separate bank account each time an interim certificate is due for payment. Reference to holding retention as a trustee is peculiar to JCT forms of contract. With regard to other standard forms such as ICE 6th and 7th Editions, GC/Works/1 1998 and the Engineering and Construction Contract (NEC) there is no obligation for employers to hold retention in a separate account.

7.5 If an employer became insolvent what liability does the contractor have for paying subcontractors who are owed money when no further sums are forthcoming from the employer?

7.5.1   When an employer becomes insolvent on construction projects there is usually substantial money outstanding to contractors. Where the JCT contract applies the contract provides for retention to be set aside in a separate bank account: *Rayack Construction Ltd* v. *Lampeter Meat Co Ltd* (1979) and see 7.3. If this has occurred the retention money which is held in trust should be paid over in full by the receiver or liquidator to the contractor. Nevertheless, even where this has happened there are usually substantial amounts due other than the retention. Subcontractors who are one down the line will often be owed sums of money and disputes arise with the main contractor as to who has taken the risk of the Employer's insolvency.

7.5.2   Contractors in the past normally included in their terms and conditions of subcontract clauses which have the intention of passing down to the subcontractor the risk of non-payment or late payment by the employer for any reason, sometimes referred to as 'pay when paid' clauses. The Housing Grants, Construction and Regeneration Act 1996 section 113 (1) (which applies to all contracts entered into after 1 May 1998) effectively outlaws pay when paid clauses but an exception is made in respect of non-payment due to the employer's insolvency. It seems clear, therefore, that if the contractor insists on a clause in the subcontract that the risk of non-payment *due to the insolvency of the employer* rests with the subcontractor then it is with the subcontractor that the loss will lie.

7.5.3   Where the standard forms of subcontract used with JCT contracts, for example DOM/1 and NSC/C have been used, then different criteria may apply. These contracts do not provide for passing the risk of insolvency down to the subcontractors. Payment is due in relation to either the issue of a payment certificate under the main contract or a payment period usually monthly.

7.5.4   Payment may even be due without the issue of such a certificate if the employer becomes insolvent. In *Scobie & McIntosh Ltd* v. *Clayton Browne Ltd* (1990), Scobie and McIntosh were nominated subcontractors for the supply and installation of catering equipment in keeping with the alternative nomination procedure under JCT 80; the subcontract conditions NSC/4 formed the basis of the subcontract.

Disputes then arose concerning payment to a number of subcontractors including Scobie and McIntosh who had two invoices outstanding.

Clayton Browne, the contractors, were faced with a battle on two fronts. They were seeking payment for their work from the employers via arbitration, whilst at the same time fending off claims for payment from their subcontractors, including Scobie and McIntosh. Their defence was that under the terms of the nominated subcontract conditions they were only obliged to pay a subcontractor within 17 days of the issue of an architect's certificate under the main contract. No such certificate had been issued by the architect, as there was a dispute with the employers proceeding to arbitration. The employer determined his contract with Clayton Bowmore Ltd who in turn determined the employment of Scobie and McIntosh Ltd.

Scobie and McIntosh argued that the normal payment arrangement did not apply as the subcontract had been determined. Their case was that they were in no way at fault and required payment from the main contractor and were not content to await the outcome of the dispute under the main contract.

Judge Davies QC was sympathetic to the subcontractor's position. He said the object of nomination was to permit employers to participate in the selection of subcontractors, but not to insulate main contractors against default by employers.

Clayton Bowmore were ordered to pay on account a sum of £57 840, which was the minimum amount due. The dispute concerning the balance would be fought out in arbitration.

Contractors are therefore very much at risk where the subcontractor is nominated under a JCT main form as they may finish up paying the full value of work carried out by a nominated subcontractor even though the architect has certified a lesser amount. This will however only apply if the subcontractor's employment has been determined through no fault of the subcontractor.

SUMMARY

The contractor's obligation to pay subcontractors if the employer becomes insolvent will be dependent upon the wording of the subcontract. DOM/1, for example, makes it clear that payment will be made irrespective of whether the employer has paid the main contractor or not. There is a court case indicating that where the NSC/C nominated subcontract form applies the contractor may be obliged to pay subcontractors even though payment has not even been certified by the architect. If the employment of the sub-contractor is determined due to no fault of the subcontractor, a contractor who wishes to pass down the risk of non-payment due to insolvency by the employer must include in the conditions of the subcontract a term to that effect.

The Housing Grants, Construction and Regeneration Act 1996, whilst outlawing pay when paid clauses, makes an exception for cases where the employer becomes insolvent.

7.6   Can a contractor/subcontractor legitimately walk off site if payment isn't made when due?

7.6.1   The law concerning a contractor/subcontractor's rights to suspend work for non-payment has been amended as a result of the Housing Grants, Construction and Regeneration Act 1996. The Act applies to contracts entered into after 1 May 1998.

7.6.2   Under the Act there is a right to suspend work for non-payment under section 112 which states:

'(1) Where a sum due under a construction contract is not paid in full by the final date for payment and no effective notice to withhold payment has been given, the person to whom the sum is due has the right (without prejudice to any other right or remedy) to suspend performance of his obligations under the contract to the party by whom payment ought to have been made (the party in default').

(2) The right may not be exercised without first giving to the party in default at least seven days' notice of intention to suspend performance, stating the ground or grounds on which it is intended to suspend performance.

(3) The right to suspend performance ceases when the party in default makes payment of the amount due in full.'

7.6.3   The standard forms have been amended to take account of the requirements of section 112 (1) of the new Act.

JCT 98 clause 30.1.4 states that the contractor may suspend work if the employer fails to pay the contractor in full and the failure continues for seven days after the contractor has given the employer, with a copy to the architect, written notice of an intention to suspend work. The contractor is entitled to continue to suspend work until payment has been made in full.

An entitlement to an extension as a result of the suspension of work arises under clause 25.4.18 with loss and expense under clause 26.2.10, provided the suspension was not frivolous or vexatious. Similar provisions exist in the revised DOM/1 under clauses 21.6 (right to suspend work), 11.10.18 (extension of time) and 13.3.10 (loss and expense).

GC/Works/1 1998, condition 52, in like manner to JCT 98, provides the contractor with a right to suspend work subject to a seven day warning notice. Condition 52 (4) provides for an extension of time but there is no right to the payment of any expense incurred as a result of the suspension.

There is no express provision concerning suspension of work for non-payment in the ICE 7th Edition or in the amendments to the 6th Edition which were issued following the implementation of the Act, or those issued in connection with the Engineering and Construction Contract (NEC). The basic requirements of section 112 (1), (2) and (3) of the Act quoted above will therefore apply.

7.6.4   Prior to the Act coming into operation and in the absence of an express clause giving a right to suspension or determination due to non-payment what were the contractor or subcontractor's rights? The decision in *DR Bradley (Cable Jointing) Ltd* v. *Jefco Mechanical Services* (1989) may be a pointer as to their entitlement.

Jefco were main contractors for the refurbishment of Islington Town Hall. Bradley was appointed as a domestic subcontractor for the electrical installation. The subcontract between Jefco and Bradley was made verbally following the submission of Bradley's tender. No specific payment terms of the subcontract were accepted but it was agreed that interim payments would be made. Bradley made various requests for payment as work proceeded but none were met in full. On 29 October 1985 Bradley made another application for an interim payment and gave notice that in the event of non-payment they would leave site. No payment was made and on 12 December 1985 Bradley left the site and litigation commenced.

It was held that the earlier underpayments by Jefco did not amount to a repudiatory breach but the non-payment of the last application was such a breach as it reasonably shattered Bradley's confidence in being paid. The court decided that Bradley were entitled to consider the contract at an end and determine.

From this decision it can be seen that, in the absence of an express clause in the contract, non-payment of interim certificates could still give rise to a contractor or subcontractor's entitlement to *determine* if such non-payment undermined the essence of the contract. However, in the absence of an express clause giving a right to *suspend* work, no such right exists at common law. The options appear to be either to carry on working or, if the non-payment 'shatters confidence', to determine the contract as the only other remedy. This is still the case for contracts outside the scope of the 1996 Act.

7.6.5   The Court of Appeal in the case of *The Channel Tunnel Group Ltd* v. *Balfour Beatty Construction Ltd and Others* (1992) (prior to the new Act) held that under English law, in the absence of an express clause in the contract, a contractor has no rights to suspend work where the employer refuses to make an interim payment in respect of part of the works.

7.6.6   The case of *C.J. Elvin Building Service Ltd* v. *Peter & Alexa Noble* (2003) involved the suspension of work. A dispute arose concerning non-payment in respect of a contract to carry out extensive work at the home of Mr and Mrs Noble. The contract was not covered by the Act as it involved undertaking work for a residential occupier. It seems at the stage where the case came before the court, C.J. Elvin was owed in the region of £40 000. Work had been stopped due to non-payment. Mr and Mrs Noble argued that the stopping of work constituted a repudiatory breach and that non-payment by an employer is not generally a breach which will enable the contractor to treat the contract as ended.

The judge held that the refusal to pay money which is due, particularly if it is a sizeable amount, or a threat not to pay any more, could amount to a repudiatory breach. In finding for the contractor, it was held that the suspension had been triggered by Mr and Mrs Noble's non-payment and could not therefore in itself be repudiatory.

SUMMARY

The law was changed by the Housing Grants, Construction and Regeneration Act 1996. Prior to the coming into effect of the Act on 1 May 1998 a contractor or subcontractor could suspend work for non-payment only where a clause was written into the contract which give that right. This, however, was changed by the Act which provides a statutory right of suspension for non-payment.

7.7 Where a contractor undertakes work which he considers should be paid for on a daywork basis and submits daywork sheets as required under the terms of the contract to the architect/engineer, if the daywork records are not signed by the architect or engineer, how does this affect the contractor's entitlement to payment?

7.7.1   Most standard forms of contract provide, under clearly defined circumstances, for extra work to be paid for on a time and materials basis, often referred to as daywork. There is usually a time-scale within which the contractor is required to submit the daywork records to the architect/engineer for approval and signing.

Contractors regularly complain that they go through the process of producing daywork records and submit them on time to the architect or engineer, only to have them left unsigned. The reasons for a failure on the part of the architect or engineer to sign the daywork records are not always clear. Sometimes it is due to forgetfulness, but it is often due to a disagreement with the information included in the daywork records.

7.7.2   The lack of properly authenticated and signed daywork records was the basis of dispute in the case of *JDM Accord* v. *The Secretary of State for the Environment, Food and Rural Affairs* (2004). During 2001 foot and mouth disease spread throughout the UK and as a result six million animals had to be slaughtered. A number of contractors, of whom JDM Accord was one, were appointed by the Secretary of State to slaughter the animals, prepare the burial sites for disposal of the animal carcasses and construct the necessary infrastructure to support the operation. The terms of the agreement between the parties provided for payment to be made on a time recorded basis. It was the intention for the Secretary of State to have a representative present on each of the sites to record the time taken by the operatives and plant. Time records were to be made out giving details of all the time spent, and these would then be signed each day by the representatives of both parties. Unfortunately the Secretary of State did not have a representative on all the sites, so JDM Accord had nobody with whom they could agree the time they recorded.

7.7.3   When it came to payment, in the absence of daywork records signed by both parties, there was a refusal to make payment in full. It was considered by Judge Thornton that, as there had been no challenge to the time recorded when the timesheets were submitted, the onus was on the Secretary of State to prove that the information included on the timesheets was incorrect. Judge Thornton held that the timesheets submitted by JDM Accord should be allowed to stand without any further proof as to their accuracy unless the Secretary of State could produce convincing evidence that the information on the timesheets was incorrect.

SUMMARY

The case of *JDM Accord* v. *The Secretary of State for the Environment, Food and Rural Affairs* (2004) makes it clear that if, having received daywork records from the contractor, the architect or engineer does not sign them, he is at risk of either having to prove at a later stage that the information included in the records is wrong or certifying them for payment in full. If the architect or engineer challenges the information included in daywork records at the time they are submitted, it is essential for full details to be provided of the hours alleged to be inaccurate. If no agreement can be reached, then the dispute resolution procedure included in the contract may have to be implemented.

7.8 Can an architect/engineer sign a daywork sheet and then refuse to certify the sums involved for payment? Is a quantity surveyor entitled to reduce the hours included on a signed daywork sheet if he considers them unreasonable or excessive?

7.8.1   A daywork sheet is a record of time, equipment and materials employed with regard to a particular on-site operation which is described therein. A signature of the architect, engineer, clerk of works or any other employer's representative merely indicates that the hours recorded and materials listed have been employed in that operation. Nothing more and nothing less. It does not indicate an intention to make payment as the work as described on the daywork sheet may be catered for elsewhere in the contract. This applies whether or not the signature is accompanied by the words 'for record purposes only'.

7.8.2   Where, however, payment on a daywork basis is established and a daywork sheet has been signed by, for example, the architect, it is not for the quantity surveyor to reduce the hours on the grounds, that he considers them excessive. If the hours are in any way incorrect then the signature should not have been appended.

7.8.3   JCT 98 clause 13.5.4 stipulates that, when valuing a variation where work cannot be properly valued by measurement, then the variation shall be calculated in accordance with the *Definition of Prime Cost of Daywork carried out under a Building Contract* issued by the RICS.

Clause 13.5 then goes on to state:

> 'Provided that in any case vouchers specifying the time daily spent upon the work, the workmen's names, the plant and the materials employed shall be delivered for verification to the Architect or his authorised representative not later than the end of the week following that in which the work has been executed.'

There is no reference to any input by the quantity surveyor and therefore once a decision has been taken to pay on a daywork basis and sheets have been submitted and signed by the architect or his authorised representative the quantity surveyor's role is of a very limited nature.

7.8.4   The ICE 6th and 7th Editions provide for payment on a daywork basis if the engineer in his opinion considers it necessary or desirable. Where a quantity surveyor is involved, in like manner to a JCT contract, he has no power to alter hours which have been agreed by the engineer or his representative.

7.8.5   In the case of *Clusky* v. *Chamberlain* (1994) it was held by the Court of Appeal that the judge in the lower court was wrong to go behind the timesheets to establish an entitlement as to quantum. It was at no time suggested that the timesheets were fake. There is no entitlement to argue that the workman did not work as expeditiously as they might.

SUMMARY

An architect or engineer may sign a daywork sheet 'for record purposes only' and then refuse to certify for payment the sum included therein.

However a quantity surveyor has no power to alter hours which he considers to be excessive on a signed daywork sheet.

7.9 Where a contractor/subcontractor includes an unrealistically low rate in the bills of quantities, can he be held to the rate if the quantities substantially increase?

7.9.1   Most quantity surveyors and engineers, if convinced that the contractor or subcontractor has included an unrealistically low rate in the bill of quantities, will insist upon the rate applying up to the quantity in the bill. Any excess in quantities over and above the bill quantity is to be paid for at a fair and reasonable rate. This seems a rational and fair approach but there are two legal cases that cast doubt upon it.

7.9.2   In the case of *Dudley Corporation* v. *Parsons & Morrin Ltd* (1959) a contract for the building of a school was let using the RIBA 1939 form, with quantities. The contract terms in issue were essentially the same as those of JCT 98. The contractors priced an item for excavating 750 cubic yards in rock at £75, i.e. two shillings a cube. This was a gross underestimate, although it was not known whether rock would be met. In carrying out the excavations described in the drawings and bills, the contractors excavated a total of 2230 cubic yards – of rock. The architect valued the work at two shillings a cube for 750 cubic yards, and the balance at £2 a cube. £2 was not unreasonable if no other price applied. The employer disputed this amount. The arbitrator found 'no sufficient evidence ... that the price of two shillings per yard cube ... was a mistake'.

Mr Justice Pearce in the Court of Appeal when deciding the matter related to the case said:

'In my view, the actual financial result should not affect one's view of the construction of the words. Naturally, one sympathises with the contractor in the circumstances, but one must assume that he chose to take the risk of greatly under-pricing an item which might not arise, whereby he lowered the tender by £1425. He may well have thought it worth while to take that risk in order to increase his chances of securing the contract.'

7.9.3   The Court of Appeal in *Henry Boot Construction* v. *Alstom Combined Cycle* (2000) gave a decision concerning measurement and rating of sheet piling. Alstom was the employer for the construction of a combined cycle gas turbine power station at Connah's Quay. Boot carried out the civils work employing an ICE 5th Edition contract. In post tender negotiations Boot submitted a price of £250 880 for additional and different temporary work in connection with the lowering of cold water pipework. This extra work formed part of the contract. In calculating this figure Boot had made an error, the result of which was that the sum was much higher than it should have been. This error remained undetected during post tender, pre-contract negotiations. A variation was issued to the Heat Recovery Steam Generator and Cooling Towers during the course of the work, involving additional sheet piling. Henry Boot in pricing the sheet piling extrapolated a rate from the price agreed for sheet piling which formed part of the £250 880 agreed in relation to lowering the cold water pipework. This rate being higher than it should have been produced an inflated rate for the varied work – a windfall to use the court's term. Alstom argued that clause 55(1)(b) of the contract provided for the use of contract rates 'as far as may be reasonable'. It was unreasonable, they said, to use a rate which was obviously included in error. Further, they argued that, as the sheet piling in the pre-contract agreed price was part of a lump sum, it was not reasonable to extrapolate a rate for sheet piling from this amount.

The Court of Appeal agreed with the judge in the lower court in finding in favour of Henry Boot. The reference in clause 55(1)(b) to the use of the rate being reasonable meant not that the rate itself had to be reasonable, but that the process of applying the rate must be reasonable. The parties had agreed the rate and they were stuck with it; the mistake was irrelevant. The Court of Appeal referred the matter back to the arbitrator for him to decide whether it was reasonable to extrapolate the rate from the £250 880 lump sum in spite of the conclusion that there was insufficient information to show how Henry Boot had calculated the sum in the first place. The process has thus gone full circle – starting with an appeal from the arbitrator's decision to the TCC on to the Court of Appeal and back to the arbitrator.

7.9.4   Both of these cases indicate that where errors occur in contract rates the rate will not be altered merely because of a substantial change in quantity.

7.9.5   Where a priced bill of quantities has been vetted by a quantity surveyor or engineer any errors which are discovered should be drawn to the contractor's attention. Any failure to do so could result in the contractor becoming entitled to an adjustment of the rate.

7.9.6   A pricing mechanism issued by the contractor resulted in *Aldi Stores Ltd v. Galliford* (2000) coming before the courts. A contract was let for the construction of a store employing the JCT IFC 84 conditions of contract. The price included for disposing of contaminated material to a licensed tip at a rate of £44.60 and the disposal of clean material at a rate of £8.50. During negotiations the contractor agreed to absorb the rates for the disposal of both contaminated and clean material into the overall price for the construction of the store. The rates for the disposal of both contaminated and clean material was thus nil. When work got underway it transpired that all the materials were contaminated and had to be disposed of at the licensed tip. The contractor claimed that this constituted a variation. No agreement was reached and the matter was referred to arbitration.

The arbitrator held that there had been an error in the bill of quantities and that the contractor was entitled to an architect's variation. A rate of £36.10 was awarded, being the difference between the rate for contaminated material and the rate for clean material. The employer appealed. It was the decision of the court that the arbitrator was wrong. There had been no change in the condition under which the work had been carried out, nor a significant change in the quantity of the work. The work had simply been moved from one category to another, i.e. from uncontaminated material to contaminated material. As the rate for both categories was nil, there was no justification in amending the price.

SUMMARY

There is legal authority for what should happen where an unrealistically low or high rate is included in a bill of quantities and the billed quantity is the subject of a substantial increase, the billed rate, whether unrealistically low or high applies.

7.10 Can a debtor enforce acceptance of a lesser sum in full and final settlement?

7.10.1 Pressure is often applied by a debtor trying to reduce the amount he owes by making an offer to pay a sum which is less than the amount due. The methods involved vary from requiring the creditor to sign an agreement 'in full and final settlement' made out in a lesser amount than the amount owed or sending a cheque for the lesser amount stated to be 'in full and final settlement'.

7.10.2 It is common procedure for contractors and subcontractors to be expected to sign a statement to the effect that they agree to payment of the balance due on the final account and claim on a 'full and final settlement' basis. Frequently the contractor or subcontractor is informed that payment will not be made unless the form is signed. The contractor or subcontractor often signs, as he needs money, even though he considers a greater entitlement is due.

7.10.3 Where one party, i.e. the contractor or subcontractor, has completed his side of the original bargain and agrees to forego an entitlement to the full payment as first agreed there must be consideration if the later agreement is to be binding. In other words the contractor or subcontractor must have received some benefit. For example, if the payment were to be made earlier than contractually required then early payment would amount to consideration for payment of the lesser sum. If the early payment is accepted the contractor or subcontractor would then have no claim to the balance.

This is sometimes referred to as 'accord and satisfaction'.

If an agreement is purportedly made whereby the contractor or subcontractor agrees to forego part of his entitlement in the absence of any such consideration the agreement will not be binding.

7.10.4 It may be, however, that the final balance offered includes a sum which is disputed. The contractor considers the amount is less than he is entitled to but the employer believes he is paying more than is due. *Hudson's Building and Engineering Contracts* 10th edition at page 22 states:

> '... consideration may be present in such a case as some bona fide dispute exists and a claim is given up in return for the promise to accept less.'

This being the case the agreement will be binding.

7.10.5 In the case of *D & C Builders Ltd* v. *Rees* (1965), an employer indicated to decorators that unless they agreed to accept a sum substantially less that the amount of their account she would pay them nothing at all. They signed a written document agreeing to accept the reduced payment in full satisfaction of their claim. Later they sued for the full amount. It was held by the Court of Appeal there was no true 'accord and satisfaction' as the plaintiffs had acted as a result of a threat which was without any justification and there was no consideration present.

7.10.6 In *Newton Moor Construction Ltd* v. *Charlton* (1981) the plaintiff contractor undertook some work for the defendant for £11 020. After the work was completed, the plaintiff sent in an invoice for £18 612. There had been some variations to the work, some by agreement and some not. As a consequence completion was delayed and the defendant claimed entitlement to set-off. He recalculated the amount he considered to be due, partly on agreed prices and partly on what he thought was a reasonable basis for the additions, deductions and delay, arriving at a figure of £8847. A cheque for this amount was sent to the plaintiff's solicitors, with an accompanying letter that this was to be regarded as being in 'full and final settlement' of any outstanding sums.

The plaintiff's solicitors responded that they were only accepting it in part payment, and issued a writ for the balance. The defendant maintained that the acceptance of the cheque amounted to accord and satisfaction.

It was held that as no part of the £8847 was disputed the sum offered by the defendant was, in effect, an admission that that amount was due. It did not matter whether the sum was termed a compromise or an accord and satisfaction, the letter, the cheque and the plaintiff's acceptance of it did not amount to accord and satisfaction without reciprocal benefit to the plaintiff.

7.10.7 In *Stour Valley Builders* v. *Stuart* (1993) it was held that a cheque sent in full and final settlement is not conclusive evidence of accord and satisfaction. The recipient of the cheque wrote indicating that he did not consider the payment as full and final, banked the cheque and successfully sued for more of the balance.

The banking of a cheque sent in full and final settlement does not stop the creditor from commencing an action for a balance claimed to be due: *Auriema Ltd* v. *Haig and Ringrose* (1988).

7.10.8 The case of *Clarke* v. *Nationwide Building Society* (1998) also deals with a cheque sent in full and final settlement.

The plaintiff wished to purchase a property and sought an advance from the defendant. A survey was carried out on instructions of the defendant, the cost of which formed part of the mortgage application fee. After the plaintiff moved in, damp appeared. The solicitor acting for the defendant was advised that the damp had been caused by condensation, a problem which could not have been evident when the survey had been carried out. The defendant considered there was no liability, but as a goodwill gesture wrote offering to refund the mortgage application fee. The offer was verbally rejected by the plaintiff, but nonetheless the defendant refunded the cheque for the mortgage fee 'in full and final settlement of the matter'. The cheque was presented and cleared. Nearly two years later, the plaintiff commenced an action against the defendant for negligence. The defendant's case was that the acceptance of the cheque amounted to accord and satisfaction which, in effect, put an end to the matter. It was held by the Court of Appeal that the wording of the defendant's letter did not amount to an offer capable of being accepted. Sending a cheque as a goodwill gesture in the hope of avoiding further complaint was not the same thing as offering a payment in settlement of all claims. The Court of Appeal found in favour of the plaintiff, who was allowed to continue the action.

7.10.9 Another case involving a cheque sent in full and final settlement is *Bracken* v. *Billinghurst* (2003). A dispute arose as a result of construction work carried out on the property of Mr Bracken and Ms Trickett. The dispute was referred to adjudication, which found in favour of Mr Bracken in the sum of £43 984. No payment was forthcoming and prior to the court action Mr Bracken rather surprisingly offered to settle for a payment of £6000. Billinghurst, however, sent off a cheque for only £5000 'on the strict understanding that the sum is offered to you in full and final settlement'. The letter went on to state that the settlement 'will be deemed to have been accepted by you, and therefore contractually binding if it is presented to the bank and cleared for payment'. The cheque was paid into the bank and cleared. Within a week Mr Bracken withdrew his offer of £6000 and demanded £38 984, being the original adjudication decision less £5000 paid.

Judge Wilcox followed the line of argument used in the *Stour Valley* case, that 'cashing a cheque is always strong evidence of acceptance (of a settlement offer) especially if it is not accompanied by immediate rejection'. The judge considered that one of the tests was whether the words and conduct of the creditor would cause the debtor, as a reasonable person, to believe the matter was final. Taking into account all the circumstances, the court held that the payment of £5000 was full and final.

7.10.10 In its decision in *Hurst Stores & Interiors Ltd* v. *ML Property Ltd* (2004), the Court of Appeal criticised construction managers Mace for allowing a subcontractor to enter into a 'full and final settlement' agreement with the client without the subcontractor understanding the implications. The subcontractor had not known that the settlement payment included variations, whilst Mace did. The court rectified the agreement for unilateral mistake, finding that Mace had acted unconscionably.

The court also found that the project manager acting for Hurst did not possess the necessary authority to enter into a full and final settlement agreement.

SUMMARY

An agreement showing a balance due on a final account which is less than the contractor's proper entitlement will only be binding as accord and satisfaction provided the contractor receives some benefit from the agreement such as payment earlier than otherwise would be the case.

If no benefit is derived then the agreement will not be binding.

However, where a bona fide dispute exists and a claim or part of a claim is given up by the employer in return for a promise to accept less on the part of the contractor, the agreement may be binding.

When a cheque is sent in full and final settlement and is banked it is advisable for the creditor to make it clear that the cheque is accepted as a payment on account but unless the cheque includes sums genuinely in dispute its acceptance will not affect the recipient's right to sue for any balance due.

7.11   How can subcontractors avoid pay when paid clauses?

7.11.1   Contractors, ever ready to pass risk down the line to subcontractors when employing non-standard conditions, usually include a clause making payment to themselves a condition precedent to payment to the subcontractors. This type of clause is often referred to as 'pay when paid'.

7.11.2   The law concerning pay when paid clauses has become subject to statutory control as a result of the Housing Grants, Construction and Regeneration Act 1996 which applies to contracts entered into after 1 May 1998. The Act seeks to outlaw pay when paid clauses except in respect of insolvency on the part of the original paying party.

Section 113(1), which could have been drafted with phraseology of a more digestible nature, states:

> 'A provision purporting to make payment under a construction contract conditional on the payer receiving payment from a third person is ineffective, unless that third person, or any other person payment by whom is under the contract (directly or indirectly) a condition of payment by that third person, is insolvent.'

7.11.3   The majority of standard forms of subcontract provide for payment to the subcontractor in accordance with the terms of the subcontract regardless of whether the main contractor has received payment from the employer. For example the nominated subcontract form NSC/C for use with JCT 98 provides for the main contractor to pay the subcontractor within 17 days of the date of issue of an interim certificate. DOM/1, the domestic subcontract for use with JCT 98, under clause 21.2 includes for interim payments at monthly intervals.

7.11.4   The exception is the CECA Blue Form of subcontract for use with the ICE conditions. Clause 15(3)(a) of these conditions entitles the subcontractor to be paid within 35 days of the specified date as set out in the first schedule. The main contractor is, however, entitled to withhold payment where the engineer has not certified in full the quantities included in the subcontractor's application or, if the engineer has certified in accordance with the subcontractor's application, the employer has neglected to make payment. This has been amended to comply with the 1996 Act and limits the contractor's rights to withhold payment due to non-payment by the employer to cases of insolvency.

7.11.5 It may be appropriate to examine some legal cases which have tried to unravel the meaning of pay when paid clauses. These cases will still be relevant in respect of clauses which give the contractor a right to refuse payment due to the insolvency of the Employer.

In *Aesco Steel Incorporated* v. *J.A. Jones Construction Company & Fidelity and Deposit Company of Maryland* (1985), an American case, Aesco entered into a subcontract with Jones to supply structural steel and metal decking for an amphitheatre. Payment was not due to be made to the subcontractor until after the owner had paid Jones. A balance of $80 320 remained unpaid to Aesco and Jones argued that as payment for this amount had not been received from the owner no payment was due to Aesco. It was held by the court that Aesco was entitled to be paid within a reasonable time even though the owner still had not paid Jones.

7.11.6 A slightly different situation arose in, another American case, *Nicholas Acoustics & Speciality Company* v. *H & M Construction* (1983). The main contract provided that the main contractor would make monthly progress payments subject to 10% retention. This retention was expressed as not being due for release until completion of the work and evidence that subcontractors had been paid. The terms of the subcontract allowed the main contractor to withhold 10% of sums due to the subcontractor until payment had been received from the owner.

Delays were encountered during the construction of the work, the owner sued the main contractor for delays and refused to pay the retention. The subcontractor sued the main contractor. Construing the payment provisions of the prime contract and subcontract together, the court said that a literal reading would result in a 'Catch 22' situation whereby the owner would never be required to pay until the subcontractors were paid, who in turn would not be paid until the main contractor was paid by the owner. It was held that the main contractor was obliged to pay the subcontractor within a reasonable time after completion of the work.

7.11.7 The New Zealand case of *Smith & Smith Glass Ltd* v. *Winstone Architectural Cladding Systems Ltd* (1991) throws an interesting light on the problem.

The appropriate wording in the sub-subcontract was:

> 'We will endeavour (this is not to be considered as a guarantee) to pay these claims within 5 days after payment to Winstone Architectural Ltd of monies claimed on behalf of the subcontractor.'

The court drew a distinction between an 'if' clause, i.e. if we are not paid you will not receive payment, and a 'when' clause, i.e. we will pay you when we have been paid. An 'if' clause makes it plain that payment will only be made after payment has been received. The 'when' clause was considered by the courts to indicate the time for payment only and that payment up the line was not a condition of payment down the line.

In the case in question the payment clause was considered by the court to be a 'when' clause and therefore non-payment of Winstone by Angus was no excuse for non-payment of Smith and Smith by Winstone.

Master Towle, the judge, considered that unless the clause spells out in clear and precise terms that payment will not be made until payment is received, the clause does no more than indicate the time for payment.

> While I accept that in certain cases it may be possible for persons contracting with each other in relation to a major building contract to include in their agreement clear and unambiguous conditions which have to be fulfilled before a sub-contractor has the right to be paid, any such agreement would have to make it clear beyond doubt that the arrangement was to be conditional and not to be merely governing the time for payment. I believe that the *contra proferentem* principle would apply to such clauses and that he who seeks to rely upon such a clause to show that there was a condition precedent before liability to pay arose at all should show that the clause relied upon contains no ambiguity.
>
> . . .
>
> For myself I believe that unless the condition precedent is spelled out in clear and precise terms and accepted by both parties, then clauses such as the two particular ones identified in these proceedings do no more than identify the time at which certain things are required to be done, and should not be extended into the 'if' category to prevent a subcontractor who has done work from being paid merely because the party with whom he contracts has not been paid by some one higher up the chain.

7.11.8   This reasoning was followed in Hong Kong case, *Wo Hing Engineering Ltd* v. *Pekko Engineers Ltd* (1998) where the words 'This contract is based on back to back basis including payment terms' were not sufficiently robust to allow the main contractor to withhold payment from the subcontractor on the grounds that money had not been received from the employer.

7.11.9   In the case of *Durabella* v. *J. Jarvis and Son Ltd* (2001), a dispute arose over a pay when paid clause. Jarvis was the main contractor on a project for the construction of a number of flats in Limehouse, London, for a developer. Durabella supplied and laid timber flooring as a subcontractor to Jarvis. The subcontract conditions included a pay when paid clause which read:

> 'Our liability for payment to you is limited to such amounts as we ourselves receive from the employer in respect of your works under this order.'

This is not so much a pay when paid clause but a limitation placed upon the sum which will be paid.

The developer was disappointed with the performance of Jarvis and terminated their contract. As is commonplace when a termination takes place, a dispute arose over money. The dispute was not resolved and the matter was referred to court. The proceedings were settled and a compromise agreement drafted. This agreement, in terms which are commonplace, stipulated that the developer was to pay Jarvis a sum of money, in this case £550 000 in full and final settlement of all claims arising out of the development. The agreement, however, included a rather unusual term to the effect that no value was included in the deal for works undertaken by Durabella.

Disputes over payment between Jarvis and Durabella had been on-going. Once the dispute with the developer was settled, Durabella no doubt thought that their dispute with Jarvis would be resolved. Jarvis, however, drew attention to the payment clause and the wording in the settlement agreement. The result was that Jarvis refused to offer any payment for the flooring work.

Durabella took their grievance to court. Their first argument, that the clause was unreasonable, left the court unmoved. In principle there is nothing to prevent a contractor passing on to subcontractors the risk of non-payment by the client, subject to the provision of the Construction Act.

The court, however, found in favour of the subcontractor. Despite the wording in the agreement to the effect that no value had been included for work undertaken by Durabella, the court was not convinced. It was considered that the wording was a smokescreen included at the behest of Jarvis to avoid paying Durabella. The court considered that Jarvis had every opportunity to ascertain how the £550 000 was made up and failed to do so. Jarvis therefore had no valid reason for refusing to make payment to Durabella.

7.11.10 The exception to the status of pay when paid clauses under the Act, which applies when the employer becomes insolvent, was the subject matter of the decision in *Hills Electrical and Mechanical plc* v. *Dawn Construction Ltd* (2003). Hills, a subcontractor, entered into a contract with Dawn which included a pay when paid clause. Part way through the subcontract work the employer became insolvent. Prior to the insolvency the pay when paid clause was inoperable. However, in accordance with the Act, contractors can avoid paying subcontractors by operating a pay when paid clause where the employer has become insolvent. The payment clause in the subcontract provided a final date for payment of 28 days after the due payment date. Had the Scheme for Construction Contracts applied, this period would have been 17 days and Hills would have been paid. Unfortunately for Hills the insolvency took place between the 17 days and 28 days. They argued that the subcontract failed to provide dates by which the subcontractor should make application for payment. This being the case, the whole of the payment provision would cease to apply and therefore the Scheme would become operative. The court did not agree as it considered that the Scheme could be used in isolated cases of non-compliance by the subcontract conditions, to fill the gaps without having to import major sections.

SUMMARY

Following the implementation of the 1996 Act contractors pay when paid clauses will be limited in their effect to situations where the employer has become insolvent. Some contractors may decide to amend their payment provisions to make it clear that pay when paid only applies when the employer is insolvent. Others may retain their pay when paid clauses leaving subcontractors to complain in that the wording of the payment provision does not comply with the 1996 Act.

If a contractor wrongly neglected to make payment as required by the 1996 Act the subcontractor would have a right to suspend work (see 7.5).

In the USA the courts seem to favour subcontractors by obliging contractors to pay within a reasonable time. In New Zealand the courts have differentiated between pay when paid clauses referred to as 'if' clauses and those termed 'when' clauses. A 'when' clause indicates merely the time for payment whereas an 'if' clause makes it plain that payment will only be made after payment has been received. Very clear wording is required if payment is to be withheld.

7.12 Once the value of a contractor/subcontractor's work has been certified and paid can it be devalued in a later certificate?

7.12.1 Contractors and subcontractors rely upon interim certificates and payments to keep their businesses going. Major difficulties can arise where work certified and paid for in the early part of a contract is later re-valued at a lower price and an adjustment made in a subsequent certificate. The contractor who has already paid a subcontractor based upon the earlier certificate may have difficulty in recovering the overpayment.

Legal text books and decisions of the courts make it clear that interim certificates and payments are in fact payments on account of the final sum which is due.

Employers are therefore entitled to adjust amounts which have already been certified and paid and in essence claw back sums which amount to overpayment.

7.12.2 *Hudson's Building and Engineering Contracts*, 11th edition, at paragraph 6.187 says:

> 'As a rule, the payments contemplated by such provisions only represent the approximate value (or a proportion of it) of the work done, and possibly also of materials delivered to the site, at the date of payment, and, in the absence of express provision, they are not conclusive or binding on the Employer as an expression of satisfaction with the quality of the work or materials. It makes no difference that they are frequently expressed to represent the value of work properly done, since such a qualification is an obvious one in any provision for payment on account.'

It also has this to say at paragraph 5.007

> 'Even though a building owner may have accepted the work so that a liability to pay the price of it arises, that will not (in the absence of a provision in the contract making the acceptance binding on the Employer) prevent the building owner from showing that the work is incomplete or badly done; he may either counterclaim or set off damages in an action by the builder, or he may pay or suffer judgment to be obtained against him for the full price and later bring a separate action for his damages...'

and again at paragraph 8.116

> 'Moreover, it should not be forgotten that waiver of a breach, or a renunciation of the right to damages, or a liability to pay for the work, will not, in general, and in

the absence of express provision, be implied from acceptance of the work by the building owner or his Architect, even in the case of patently defective work.'

7.12.3   In *Fairclough Building* v. *Rhuddlan Borough Council* (1985) a nominated sub-contractor, Gunite, had its employment determined. The architect issued a renomination requiring the work to be finished off by Mulcaster. Gunite had been paid for work which was subsequently shown to be defective.

In a subsequent certificate the amount certified in favour of Gunite was reduced. By that time Gunite was insolvent.

It was held by the court that the employer was entitled to be credited with the reduced value of work carried out by Gunite. The loss therefore fell upon the main contractor's shoulders.

7.12.4   An exception to this rule applies under the ICE 6th and 7th Editions where payment has been certified and paid to a nominated subcontractor. Clause 60(8)(a) and (b) states that the engineer will not delete or reduce any sum previously certified and paid to a nominated subcontractor.

A further exception seems to occur in clause 8.2.2. of JCT 98 which states:

'In respect of any materials, goods and workmanship, as comprised in executed work, which are to be to the reasonable satisfaction of the Architect in accordance with clause 2.1, the Architect shall express dissatisfaction within a reasonable time from the execution of the unsatisfactory work.'

Clause 2.1 deals with the quality of materials or standards of workmanship which in accordance with the contract are a matter for the opinion of the architect. This clause provides that such quality and standards are to be to the reasonable satisfaction of the architect.

Where this clause applies architect and employer may have some difficulty if work is certified and paid for and much later they try to reduce the valuation of the work certified. Whether it will be possible to reduce the valuation of work once certified will depend upon the circumstances, but this will put the architect under pressure to reject substandard work early.

7.12.5   In contrast the Engineering and Construction Contract (NEC) under clause 50.5 provides for the project manager to correct any wrongly assessed amount in a later certificate.

SUMMARY

Payment of an interim certificate represents a payment on account of the final sum due. It is always open to the architect or engineer to certify a sum and later reduce the amount certified in respect of work executed.

There are exceptions, for example sums certified in respect of nominated subcontractor's work under ICE 6th and 7th Editions.

7.13 Can a contractor deduct claims for overpayments levied on one contract from monies due on another in respect of a subcontractor's work?

7.13.1   It is quite common for a contractor who has an indisputable debt due to a subcontractor to refuse to make payment on the grounds that legitimate claims have been levied or overpayments have been made on another contract or contracts which are greater than the sum due. Hence no payment is made.

7.13.2   This situation arose in the Court of Appeal case of *B. Hargreaves Ltd* v. *Action 2000* (1992). In this case 12 subcontracts were entered into between the same main contractor and subcontractor. The subcontractor was due to be paid £104 160 in accordance with three interim certificates. No sum was paid as the contractor argued that on some of the contracts overpayments had been made which exceeded the amount of the certificates. Therefore, it said, it was entitled in equity to set-off.

The court held that for the contractor to set off the overpayments, the other contracts must be closely and inseparably connected to the one where the certificates were due.

In giving judgment Judge Fox-Andrews QC stated:

> The law was restated by the Court of Appeal in *Dole Dried Fruit & Nut Co Ltd* v. *Trustin Kerwood* (1990). In his judgment Lord Justice Lloyd with which Lord Justice Beldam agreed, said:
>
> > 'But for all ordinary purposes, the modern law of equitable set-off is to be taken as accurately stated by the Court of Appeal in *Hanak* v. *Green*. It is not enough that the counterclaim is "in some way related to the transaction which gives rise to the claim". It must be "so closely connected with the plaintiff's demand that it would be manifestly unjust to allow him to enforce payment without taking into account the cross claim".'
>
> The other two subcontracts were made between the same parties, on the same day with the same building owner as the Irlam subcontract. Each of these three subcontracts related to the construction of a petrol station. But Action was under no contractual obligation to Petrofina to have entered into each of these three subcontracts with Hargreaves. Taking all circumstances into account the equitable set-off plea fails.

7.13.3   A second argument used by the main contractor was that, as the overpayments amounted to debts, they had a legal right to set-off. The court held that a claim of mutual debts was only available when the claims on both sides were in respect of liquidated debts or money demands which were readily and without difficulty ascertained. The figures sought to be set off were assessed by Action's surveyor and it was not suggested that Hargreaves accepted the figure. Without a full hearing the amount due, if any, could not be ascertained.

7.13.4   Contractors often overcome the difficulties of setting off overpayments and claims arising on one contract from monies due on another by including an express right in the terms of the subcontract.

The main contract GC/Works/1 1998 gives the employer such rights of set off where, in condition 51, it states in rather long winded wording:

'Without prejudice and in addition to any other rights and remedies of the Employer, whenever under or in respect of the Contract, or under or in respect of any other Contract between the Contractor or any other member of the Contractor's Group and the Employer or any other member of the Employer's Group, any sum of money shall be recoverable from or payable by the contractor or any other member of the Contractor's Group by or to the Employer or any other member of the Employer's Group, it may be deducted by the Employer from any sum or sums then due or which at any time thereafter may become due to the Contractor or any other member of the Contractor's Group under or in respect of the Contract, or under or in respect of any other contract between the contractor or any other member of the Contractor's Group and the Employer or any other member of the Employer's Group. Without prejudice and in addition to any other rights and remedies of the Contractor, each member of the Contractor's Group shall have rights reciprocal to those of each member of the Employer's Group under this Condition.'

SUMMARY

It would seem for a contractor to be able to set off from monies due on one contract overpayments or claims on another, he must be able to demonstrate one or more of the following

- An equitable right on the basis that the contracts in question are closely and inseparably connected (for example, separate phases of a development which have been let as separate contracts)
- A legal right to set off mutual debts which must be liquidated and not contested (an unlikely contingency in the field of contested claims)
- A clause in the subcontract which gives the contractor the necessary powers of set-off.

7.14 When a contractor completes significantly early may the architect/engineer legitimately delay certification to match the employer's ability to pay from available cashflow?

7.14.1 Most standard forms provide for the contractor completing early.

- JCT 98 by clause 23.1.1 states that the contractor 'shall complete the same [the Works] on or before the completion date'.
- ICE 6th and 7th Editions by clause 43 require the contractor to complete the work 'within the time as stated in the appendix'.
- The Engineering and Construction Contract (NEC) by clause 30.1 states that 'completion is on or before the completion date'.
- GC/Works/1 1998 in condition 34(1) refers to completion 'by the Date or Dates for Completion'.

It is clear from these clauses that the contractor is entitled to finish early.

7.14.2   With regard to the issue of the completion certificate the architect or engineer has no option but to issue one once the contractor has finished.

- ICE 6th and 7th Editions by clause 48 require the engineer to issue a certificate of substantial completion when the works are substantially completed.
- JCT 98 in clause 17.1 provides for the architect to issue a certificate of practical completion when in his opinion practical completion for the purposes of the contract has taken place.
- GC/Works/1 1998 condition 39 states: The PM shall certify the date when the Works or any Section ... are completed'.
- The Engineering and Construction Contract (NEC) by clause 30.2 states 'The PM certifies completion within one week of completion'.

7.14.3   With regard to payment the standard forms in regular use normally, provide for certification and payment of the value of the works executed on a monthly basis.

Certificates and payments must therefore be made on this basis whether or not the contractor is ahead or behind programme and even if he is likely to finish early or has indeed already completed.

For example:

- JCT 98 clause 30.2
- ICE 6th and 7th Editions clause 48
- GC/Works/1 1998 condition 47
- Engineering and Construction Contract clause 50

All provide for certification and payment of work properly carried out at the date of the certificate or other method of stating the value of work.

Some standard forms such as GC/Works/1 1998 and JCT 98 provide for milestone or stage payments.

None of these provisions relate in any way to the employer's ability to pay. The advantage of provisions for milestone or stage payments is that the employer is able to manage his cashflow in a more structured manner.

FIDIC 4th Edition in clause 14.3 tries to address this problem by requiring the contractor to produce a cash flow forecast. This allows the employer to plan his cash flow in advance on the assumption that a programme is produced by the contractor to complement the stage payment schedule and that the contractor adheres to his programme.

SUMMARY

Most standard forms of contract provide for completion of work on or before the completion date written into the contract. The payment provisions require certification and payment of work properly executed. There is not any facility for the architect or engineer to reduce the value of amounts to be certified or delay certification to suit the employer's ability to pay.

7.15 Is an employer entitled to pay a nominated subcontractor direct when the main contractor is insolvent?

7.15.1   Most standard forms of contract provide the employer with an option to pay nominated subcontractors direct. The standard forms usually limit that option to where the main contractor has failed to pay the nominated subcontractor.

The general consensus of opinion is that when a contractor is insolvent payment direct by an employer to a nominated subcontractor is unwise from an employer's point of view in that it is considered to breach the insolvency rules.

Support for this view is derived from *British Eagle International Airlines Ltd* v. *Compagnie Nationale Air France* (1973). The case involved a number of airlines who together created a clearing house for settling accounts. Members rendering services to other members would submit accounts to the clearing house after giving credit for services received from the other members. They would then either receive the balance if in credit or make payment if in debit. One airline went into voluntary liquidation. It was held that to comply with insolvency rules individual debtors were liable in full leaving individual creditors to prove in the liquidation.

In the Hong Kong case of *Right Time Construction Co Ltd (in liquidation)* (1990) it was held that direct payment contravened the insolvency rules.

7.15.2   Further support was provided by the case *Attorney General of Singapore* v. *Joo Yee Construction Pte Ltd (in liquidation)* (1992).

A main contractor entered into a contract with the Singapore Government for the construction of a Blood Transfusion Services Department. The first, second, third and fourth defendants were subcontractors on the project. The contractor went into liquidation having been paid all sums due under interim certificates but having failed to pay the nominated subcontractors in respect of two of the later certificates. The government proposed paying the subcontractors directly and deducting the sums from amounts due to the main contractor. The liquidator sought judgment on whether such direct payments would be in contravention of the Singapore Companies Act 1985 on the basis that as they would amount to preferential payments to creditors and prevent a *pari passu* settlement of the company's liabilities which was required by the Act.

Under clause 20(e) of the contract, which was similar to the JCT provisions, the employer was entitled to pay nominated subcontractors directly and to deduct such sums from future certificates. An additional power to do so existed in the event of a winding up.

It was held that a payment under clause 20(e) would contravene sections 280(1) and 327(2) of the Singapore Companies Act 1985. The liquidator was not bound by the clause, and any payment made to subcontractors would be void against him.

7.15.3   A similar decision arose in the Northern Ireland case of *B. Mullan & Sons (Contractors) Ltd* v. *Ross and Another* (1995). On 6 August 1993 the London-derry Port and Harbour Commissioners, the employers, entered into a contract using JCT With Contractor's Design, McLaughlin & Harvey plc

were the main contractors and B Mullan and Sons was engaged as a subcontractor to carry out the site and ancillary works forming part of the
contract. The subcontractor carried out works to the value of approximately
£200 000 in respect of which an interim payment of about half was made by
the contractor. The balance due was eventually agreed. The contractor went
into receivership on 7 October 1993, and on 26 January 1994 a resolution was
passed for voluntary winding up. The respondents were appointed as joint
liquidators.

In the autumn of 1993 the subcontractor had made application for direct
payment from the employers of the balance due, but the employers,
although willing to do so, felt unable to make payment until they had full
details of the amounts owed to all subcontractors. The liquidation occurred
before any payment could be made, and the employers did not pay the
balance either to the plaintiff or to the liquidators.

7.15.4 It was held by the Court of Appeal that the employers were not entitled to
make payment to the subcontractors. The monies should be paid to the
liquidators and form part of the general fund applicable for payment of a
dividend *pari passu* to all the company's creditors under article 93 of the
Insolvency (Northern Ireland) Order 1989, 'property' being defined in article
2 thereof. The power to make direct payment to the subcontractor was
inconsistent with this fundamental principle of insolvency law.

7.15.5 A different light has been thrown on this matter by the decision of the High
Court of Ireland in *Glow Heating Ltd* v. *Eastern Health Board and Another*
(1988).

A main contractor was engaged to build a health centre, with the
mechanical services to be undertaken by a nominated subcontractor. Under
the main contract the subcontractor's money was to be certified by the
architect for inclusion in payments made to the main contractor. The main
contractor was then to pay the nominated subcontractor, subject to any
deductions for retention money and discount. The subcontractor's retention
money was to be placed in a trust fund and the main contractor's interest in
the retention was to be held by the Employer as trustee. If the main contractor failed to pay the subcontractor, the contract provided that the subcontractor could apply directly to the employer. The main contractor went
into liquidation and £6617 had been certified as due to the subcontractor and
paid but the main contractor had not passed it on.

The parties accepted that the subcontractor was entitled to £12 148 from
the liquidator in respect of money for work carried out after the date of the
main contractor's liquidation and retention money held by the employer for
works undertaken before and after the liquidation date.

The subcontractor sought direct payment of the £6617 from the employer,
claiming that the employer could deduct the amount from retention which
would otherwise be due to the main contractor. Whilst the liquidator
accepted the validity of the trust clauses, he argued that the direct payment
provisions were void as contrary to public policy since they contravened the
local Companies Act 1963, section 275. The liquidator further relied upon
clause 22 of the subcontract which provided that, if the main contract provisions were inconsistent with the subcontract, the subcontract provisions

would take precedence and claimed that, since the direct payment provision in the main contract was expressed in a mandatory way, it was not consistent with the provision in the subcontract which was expressed in a permissive form and was, therefore, invalid.

It was held:

- The direct payment clauses in the contracts did not reduce the insolvent main contractor's property in contravention of the Companies Act since the liquidator took the property, i.e. the retention, subject to such liabilities as affected it while it was in the main contractor's hands.
- The direct payment clause in the main contract was not inconsistent with the subcontract provision.
- The employer was to pay the liquidator the sum of £12148 which was to be passed on to the subcontractor without deductions and the employer was to pay the sum of £6617 directly to the subcontractor and deduct the amount from the main contractor's retention.

7.15.6   The Court of Appeal in the *Mullan* case distinguished the decision in *Glow Heating* by saying that the employer only had an option to pay the subcontractor direct in *Mullan* but had an obligation to pay the subcontractor in *Glow Heating*.

7.15.7   JCT 98 addresses the matter in clause 35.13.5.3.4 where it states that any right for the employer to pay a nominated subcontractor direct shall cease to have effect when there is in essence

'either a Petition which has been presented to the Court for the winding up of the Contractor
or a resolution properly passed for the winding up of the Contractor other than for the purposes of amalgamation or reconstruction'

7.15.8   Agreements may be entered into whereby an employer undertakes to pay a subcontractor direct. Where this arrangement occurs, the employer takes the risk of having to duplicate the direct payments. In the case of *Brican Fabrications Ltd* v. *Merchant City Developments Ltd* (2003), Merchant entered into a contract with Circle Construction to construct a leisure centre in Glasgow. Circle was experiencing financial difficulties and as a result could not arrange a subcontract for the steel fabrication works. The preferred steelwork specialist, Brican, would only enter into a subcontract if payment was made direct by Merchant. An agreement was therefore entered into for direct payments to be made by Merchant to Brican. All went well in respect of the first six payments, and then Circle became insolvent and went out of business. Brican calculated that they were owed £43000 as a balance due on the final account. Merchant refused to make payment on the basis that the agreement was to pay to Brican money deducted from sums due to Circle, and as a result of the insolvency nothing was due. The Inner House of the Scottish Court of Sessions disagreed. The agreement to pay direct was clear and was not affected by the liquidation of Circle. This payment provision seems to run contrary to the insolvency law, leaving Merchant open to the possibility of having to pay the £43000 a second time to Circle's liquidator.

7.15.9   The case of *Action Strength* v. *International Glass Engineering* (2001) dealt with the situation where post contract an agreement is entered into whereby the employer pays a subcontractor direct. Action Strength undertook to supply labour to International Glass Engineering in connection with the construction of a float glass factory. Saint-Gobain was the employer who entered into a contract with International Glass Engineering for the work to be carried out. International Glass Engineering got into financial difficulties and a stage was reached where Action Strength was owed £200 000. They complained to the employer, Saint-Gobain, and an agreement was reached whereby Action Strength would be paid direct by Saint-Gobain. Further comfort was given to Action Strength in that Saint-Gobain held a performance bond in the sum of £700 000, and they were told this could be utilised to pay them if all else failed. The amount of money owed to Action Strength steadily increased until it reached £1.3m at which point Action Strength withdrew its labour. International Glass Engineering was insolvent and could not pay up. An action was commenced against Saint-Gobain to enforce the promise to pay Action Strength direct.

The court had to decide whether Saint-Gobain had undertaken to pay Action Strength out of its own money or merely to redirect money due to International Glass Engineering. If the answer was that Saint-Gobain had agreed to pay out of its own money following default on the part of International Glass Engineering, it would be regarded as a guarantee and under the provisions of The Statute of Frauds Act 1677, to be enforceable it must be in writing. The court held the agreement to be a guarantee and, in the absence of anything in writing, it was unenforceable. This was most unfortunate for Action Strength as they had carried on working in good faith and, based upon an undertaking from Saint-Gobain, had reasonably expected to be paid.

SUMMARY

There is now conflicting case law on this matter. A decision in Hong Kong in the *Right Time Construction Co* case held that direct payment contravened the insolvency rules, as did the Northern Ireland case of *B. Mullan & Sons (Contractors) Ltd* v. *Ross and Another* (1995) and the Singapore case of *Attorney General of Singapore* v. *Joo Yee Construction* (1992). In *Glow Heating*, an Irish case, the court took a different view. In England we have the *British Eagle* case but this decision has nothing to do with construction contracts or nominated subcontractors.

It would seem likely that the English courts in the light of the *British Eagle* and the *Mullan* cases would hold that direct payment is contrary to the insolvency rules. JCT 98 deals with the matter stating that the right to pay direct ceases when a petition to wind up the contractor has been presented to the court.

7.16 Where an architect/engineer undercertifies, is the contractor/subcontractor entitled to claim interest?

7.16.1 Contractors frequently allege that engineers and architects deliberately undervalue work in interim certificates. The tendency has been to ensure that contractors are not overpaid where there is a recession in the construction industry resulting in many contractors becoming insolvent. When under certification occurs there will often be a catching up of payments after practical completion with contractors being paid substantial sums long after the work has been completed. The amounts involved can be much greater where genuine disputes occur which are ultimately resolved. Contractors often claim interest on these late certifications but rarely receive payment.

7.16.2 A great deal of the law as it applies in this country is judge made. When judges fail to agree confusion and additional cost is inevitably the result.

Take, for example, the alternative ways in which the courts have interpreted clause 60(6) of the ICE 5th Edition. This clause states:

> 'In the event of failure by the Engineer to certify or the Employer to make payment in accordance with sub-clauses (2), (3) and (5) of this clause the Employer shall pay to the contractor interest on any payment overdue.'

7.16.3 Different judges in recent times have placed more than one interpretation as to what is meant by 'failure by the Engineer to certify'. In the Scottish case of *Nash Dredging* v. *Kestrel Maritime* (1987), Lord Ross stated:

> Accordingly if it appeared at the end of the day that the sum certified by the engineer was less than ought to have been certified in my opinion the engineer could not be said to have failed to have certified, provided that it had been his honest opinion that the sum certified by him was the amount then due.

In other words there would be no failure to certify if the engineer, for example, issued a certificate concerning, say, a claim under clause 12 and, having given the matter further consideration, later increased the amount certified – provided he acted in good faith. This decision was clear and free from ambiguity.

7.16.4 The second case to deal with an interpretation of clause 60(6) is *Hall and Tawse* v. *Strathclyde Regional Council* (1990), another Scottish decision. In this case the judge followed *Nash Dredging* using the following wording:

> I agree with Lord Ross that there would not be a failure on the part of the engineer to certify merely because the sum certified turned out to be less than the sum which the court or arbiter thought was due.

The judge went on to suggest a further situation which may call for interpretation under clause 60(6) but declined to express a view:

> 'It is not necessary for present purposes to consider whether there would be a failure on his [the engineer's] part if he had proceeded on an interpretation of the contract or some other point of law relating to matters on which his opinion is required in order to provide a certificate which is later found to be erroneous.'

7.16.5 A contrasting view was taken in England when an arbitration award was the subject of an appeal in *Morgan Grenfell Ltd and Sunderland Borough Council* v. *Seven Seas Dredging Ltd* (1990).

The matter was referred to Judge Newey. His view was:

> If the engineer certifies an amount which is less than it should have been, the contractor is deprived of money on which he could have earned money . . . If the arbitrator revises his [the Engineer's] certificate so as to increase the amount, it follows that the engineer has failed to certify the right amount.

Judge Newey upheld the decision of the arbitrator in holding that interest would be payable under clause 60(6) if the engineer, even acting in a bona fide manner, under-certified. This is in stark contrast to the Scottish decisions which deprived the contractor of a right to interest if the engineer acts honestly.

There was no appeal against Judge Newey's decision which must have sent shockwaves through establishments employing civil engineering contractors in England and Wales.

7.16.6   The Commercial Court differed from Judge Newey's decision in *The Secretary of State for Transport* v. *Birse Farr Joint Venture* (1992). The case arose out of a contract to construct part of the M25 which involved the use of the ICE 5th Edition.

The arbitrator found in favour of the contractor with regard to interest which was to be computed from a date three months after each valuation date and an appeal was lodged by the Secretary of State for Transport. Mr Justice Hobhouse in the Commercial Court, impressed by the views expressed by Mr Justice Buckley in the case of *Farr* v. *Ministry of Transport* (1960), commented as follows:

> A distinction clearly emerges from this case between the issue of a certificate which bona fide assesses the value of the work done at a lower figure than that claimed by the contractor and a certificate which, because it adopts some mistaken principle or some errors of law, presumably in relation to the correct understanding of the contract between the parties, produces an under-certification.

Mr Justice Hobhouse, in finding in favour of the Secretary of State for Transport, said that interest under clause 60(6) would only be due where the engineer undercertifies due to some mistaken principle or some error of law. A certificate which bona fide assesses the value of the work done at a lower figure than is due to the contractor and which does not involve a contractual error or misconduct of the engineer will not rank for interest under clause 60(6). This shows a marked difference from what was said by Judge Newey in the *Morgan Grenfell* case when allowing interest on under-certification even where the engineer acted honestly without making any contractual errors.

7.16.7   Another case on the subject is *Kingston upon Thames* v. *Amec Civil Engineering Ltd* (1993) where the court, on hearing an appeal from an arbitration, followed the line of the decision in *Birse Farr*.

7.16.8   In *BP Chemicals* v. *Kingdom Engineering* (1994) the ICE 5th Edition applied with clause 60(6) deleted. It was held that the arbitrator could only award interest from the date of his award.

7.16.9 The ICE 6th and 7th Editions in clause 60(7) have expanded the wording of clause 60(6) of the 5th Edition making it clear that interest is payable on under-certification and thus following the *Morgan Grenfell* decision.

The clause 60(7) wording states:

> 'In the event of
> (a) failure by the Engineer to certify or the Employer to make payment in accordance with sub-clauses (2) (4) or (6) of this Clause or
> (b) any finding of an arbitrator to such effect
> the Employer shall pay to the Contractor interest compounded monthly for each day on which any payment is overdue or which should have been certified and paid at a rate equivalent to 2% per annum above the base lending rate of the bank specified in the Appendix to the Form of Tender.
>
> If in an arbitration pursuant to Clause 66 the arbitrator holds that any sum or additional sum should have been certified by a particular date in accordance with the aforementioned sub-clauses but was not so certified this shall be regarded for the purposes of this sub-clause as a failure to certify such sum or additional sum.'

7.16.10 In the case of *Amec Building* v. *Cadmus Investment Co Ltd* (1996) a contractor's claim arising out of a JCT form of contract included interest for under-certification. The arbitrator did not follow the decision in *BP Chemicals* v. *Kingdom Engineering* (1994). He awarded interest from the date of under-certification and the employer appealed. In finding for the contractor the judge said:

> I respectfully concur in the reasoning of the authors of the Building Law Reports at page 116E, that the party seeking to review a certificate had a cause of action at the date of the under-certification and not, as Judge Harvey held, only upon the publication of the award of the arbitrator. I accept the argument that if monies previously unpaid to a contractor are subsequently found to be due by reason of the determination in the arbitration, that an award of interest should be made to compensate the contractor for the period during which such monies have been withheld from him. It would also remove the benefit, unjustly obtained as a result of the arbitrator's award, which would accrue to the employer by withholding sums which were properly due to the contractor. For all these reasons it seems to me that the arbitrator's award of simple interest in this case was perfectly proper and I therefore dismiss the head of appeal.

7.16.11 Interest claimed in respect of money set off during a period of insolvency was the subject matter in the case of *Charles Brand Ltd* v. *Orkney Islands Council* (2001). The contract, under which the ICE 5th Edition conditions applied, was for various works relating to new ferry terminals at Loth and Rapness in the Orkney Isles.

A sum of £280 880 had been certified by the engineer in three certificates, but not paid by the employer. It was the employer's case that it was owed money by the contractor in respect of a contract to supply stone and other services. The employer argued that it was entitled to a set-off as the contractor was insolvent. Subsequently the contractor received an injection of working capital from its parent company. It was agreed by all concerned that, as the contractor was no longer insolvent, the money which had been set off should be and was in fact repaid.

The contractor lodged a claim for interest. A right of set-off existed whilst the contractor was insolvent, but this came to an end following the injection of capital. Clause 60(6) of the contract obliges the employer to pay interest to the contractor should payment not be made in accordance with the terms of the contract. It was the contractor's case that the employer had failed to make payment as required by the contract and therefore payment of interest was due. Understandably the employer argued that, due to the contractor's insolvency, a right of set-off arose and therefore there was no question of a failure to make payment.

Rather surprisingly there seem to be few cases which give much guidance as to the meaning of the wording 'failure to make payment'. The court took the view that the employer was entitled to withhold payment of the principle sum to provide it with security for a debt under a separate contract. It was not a matter of withholding money due to a breach of contract. The interest would, under the circumstances, continue to accrue and as the employer had the use of the money which was withheld, the interest should be repaid to the contractor.

SUMMARY

The situation with regard to interest on under-certification had been extremely unclear due to conflicting legal decisions where the ICE 5th Edition applied and only a failure to certify at all or under-certification due to influence by the employer or a misunderstanding of the contract by the engineer gave an entitlement to interest. Interest will, however, be recoverable under ICE 6th and 7th Editions even with regard to under-certification in good faith.

Where the JCT conditions apply courts seem to take the view that an arbitrator has power to award interest from the date the certificate should have been issued. The contract does not make specific provision but the more recent cases take a view that it is equitable to allow the contractor compensation for not having use of the money during the whole period.

7.17   Can an architect/engineer refuse to include an amount of money in a certificate in respect of materials stored on site if the contractor or subcontractor cannot prove he has good title to the materials?

7.17.1   Suppliers in the recent past have been at risk after having delivered materials to site of not receiving payment. From time to time a contractor may become insolvent without having paid the supplier for his materials. The supplier will often lose out as the liquidator or receiver will take the benefit of the materials and pay out to the supplier only a small fraction of the invoiced price for the materials and often nothing at all.

Many suppliers set out to protect themselves from this type of risk by including in their terms of trading what has become known as a 'retention of title clause'. The purpose of this clause is to enable the supplier to retain ownership until he has been paid. If the supplier delivered the materials to site and the contractor becomes insolvent before paying him, the goods can be reclaimed, or payment in full demanded from the receiver, employer or whoever wishes to make use of them – provided the retention of title clause is effective.

The court held in *Hendy Lennox (Industrial Engineers) Ltd* v. *Grahame Puttick Ltd* (1983) that the following represented an effective retention of title clause.

'Unless the company shall otherwise specify in writing all goods sold by the company to the purchaser shall be and remain the property of the company until the full purchase price thereof shall be paid to the company.'

7.17.2 Employers are often concerned that they may pay for materials stored on site and subsequently the contractor becomes insolvent before paying sub-contractors and suppliers. If the unpaid subcontractors or suppliers have effective retention of title clauses in their contracts with the contractor they can demand payment from the employer if he intends to use the materials. Under these circumstances the employer may finish up paying twice for the materials as happened in *Dawber Williamson Roofing Ltd* v. *Humberside County Council* (1979).

7.17.3 Architects, engineers and quantity surveyors often consider they have a duty to protect the employer from the risk of having to pay twice for materials delivered to site. Some therefore insist upon the contractor producing proof of title before agreeing to include materials delivered to site but unfixed in an interim certificate.

7.17.4 The standard forms in common use expressly provide for payment in respect of materials delivered to site but unfixed: ICE 6th and 7th Editions (clause 60(1)(b)), JCT 98 (clause 30.2.1.2) and GC/Works/1 1998 (condition 48).

None of these clauses make any reference to certifying amounts only in respect of materials for which the contractor is able to demonstrate that he holds good title. It would be necessary for a special clause to be included which limits the contractor's entitlement to payment for materials on site to those which he actually owns. Such a clause would, however, be almost impossible to apply. The contractor for example could pay his supplier only to discover later that a sub-supplier (who had not been paid by the supplier) had an effective retention of title clause in his terms of trading. Under these circumstances the employer might not be protected.

7.17.5 It should be noted that once materials have been built into the structure of the building the retention of title clause will no longer be effective.

SUMMARY

An employer who pays for materials delivered to site but unfixed may find that he finishes up paying twice where the main contractor becomes insolvent, if the supplier has an effective retention of title clause in his terms of trading. Attempts can be made to avoid this but it seems clear that the standard forms of contract place this type of risk onto the employer.

The contracts in regular use, e.g. ICE 6th and 7th Editions JCT 98, GC/ Works/1 1998 and the like, all include for payment in interim certificates of unfixed materials on site. No mention is made in any of these contracts to proof of title being demonstrated by the contractor as a condition precedent to payment. Quantity surveyors, architects and engineers may decide to warn employers of the possible shortcomings of this type of payment clause but are not entitled to exclude unfixed materials on site from interim certificates solely on the grounds that good title cannot be proved. It would require a specially drafted clause in the contract fully to protect the employer, and even this may be difficult to implement. (See also 7.2)

7.18 Can an employer refuse to honour an architect/engineer's certificate and so delay payment?

7.18.1 It had been traditional for employers to honour the architect/engineer's payment certificate without question. The first chink in the armour appeared in the case of *CM Pillings & Co Ltd* v. *Kent Investments* (1985) where an employer refused to honour an architect's certificate and he convinced the court that there was a bona fide dispute as to its accuracy. The court ordered a stay of the application for summary judgment for the sum certified so that the matter could be referred to arbitration.

7.18.2 Another decision dealing with the same principle is *John Mowlem & Co Plc* v. *Carlton Gate Development Ltd* (1990). John Mowlem was the contractor and Carlton Gate Development the employer under the terms of the contract which incorporated a non standard form. For contractual and administrative reasons the development was divided into sections.

The architect issued two interim certificates which totalled about £2.25 m. On the last date on which payment was due, the architect sent to John Mowlem letters which purported to be notices under the conditions of contract authorising deductions from certified interim payments in the case of delays in completion of any one section of the work.

In addition Carlton Gate raised various counterclaims and, relying on the architect's purported notices and their alleged rights of set-off in respect of the counterclaims, they deducted half the sum certified by the architect before making payment.

John Mowlem contested Carlton Gate's right of deduction on the following grounds:

- The contents of the letters and the surrounding circumstances showed that they were not written in good faith.
- The counterclaims arose after the date when money due on the certificates should have been paid, and for that reason as a matter of law could not be set off against the liability to pay on the interim certificates.
- The letters written by the architect did not satisfy the requirements of the contract.

John Mowlem applied for summary judgment under Order 14 on the grounds that as there was no dispute as to the debt and the court should therefore order that payment be made. The employer contested the application requesting a stay of the court proceedings for the matter to be referred to arbitration.

7.18.3   Judge Bowsher decided in favour of the employer. He referred to the case of *Home & Overseas Insurance* v. *Mentor Insurance* (1989) where it was said that the purpose of Order 14 is to enable a plaintiff to obtain a quick judgment where there is plainly no defence to the claim.

Judge Bowsher when giving his decision had in mind the interests of other litigants who might be delayed by a lengthy hearing and the extreme pressure on the court's time referring to *British Holdings Plc* v. *Quadrex* (1989) where the judge considered applications for summary judgment inappropriate with facts so complex that days were required for the hearing and a huge weight of evidence was necessary in order to understand the issues.

However Judge Bowsher did comment that where construction contracts provide for interim payment it is usually for the very good reason that the contractors need money to continue with the project. The sums involved, he thought, were so large that even the bigger construction companies feel the pinch when payment is withheld. It was his view that in appropriate cases the court should not shrink from dealing with an Order 14 summons even if the evidence is bulky. However if the inappropriate cases can be discouraged, those cases in which it is appropriate to give summary relief will receive earlier attention.

In considering the three points put forward by John Mowlem the judge held as follows:

- It was inappropriate at a hearing for an Order 14 application to consider whether an architect had acted in bad faith.
- The contention that the employer's counterclaims due to their timing were invalid was a fundamental point of law which should be decided only after due deliberation and would therefore not be a matter for an Order 14 summons.
- The claim that the architect's letter concerning deduction from certificates did not comply with the contract could not stand on its own.

The judge refused to grant John Mowlem's application for summary judgment and granted a stay for the matter to be referred to arbitration.

7.18.4   In the case of *RM Douglas Ltd* v. *Bass Leisure Ltd* (1991) a dispute arose out of a contract which incorporated the standard JCT Management Contract 1987. The contractor applied for summary judgment in the sum of £1.3 m due on interim certificates. Bass Leisure applied for the action to be stayed and the matter referred to arbitration.

The defendant's reasons for not honouring the certificates were twofold:

- It was contended that there were grounds for questioning whether the sums certified were in fact due under the interim certificates.
- It was argued that there was an entitlement to set off damages in respect of delay and other alleged breaches of contract.

Judge Bowsher listened to arguments as to what constitutes a dispute to be referred to arbitration. He was influenced by the decision of Judge Savill in *Hayter* v. *Nelson and Home Insurance* (1990), where Judge Savill interpreted the words 'there is not in fact any dispute' as meaning the same as 'there is not in fact anything disputable'. Judge Bowsher went even further to conclude that the defendant, to defeat an application for summary judgment, need only 'in good faith and on reasonable material raise arguable contentions' as to whether sums certified are due.

It can readily be seen that if a defendant to an application has to demonstrate that there is a bona fide dispute this can be a far more taxing matter than merely showing that a proposition is disputable or merely raises contentions.

Judge Savill had summed up the situation in the following terms:

> Only in the simplest and clearest cases, that is where it is readily and immediately demonstrable that the respondent has no good grounds at all for disputing the claim, should that party be deprived of his contractual rights to arbitrate'.

In the main Judge Bowsher found in favour of the employer. He considered there were reasonable grounds for challenging the certificate with the exception of £236,253.20 which he ordered to be paid to the contractor, the balance being referred to arbitration.

7.18.5 These two decisions run contrary to the way in which the industry had operated in times past when Lord Denning once said that an architect's certificate was like a bill of exchange, i.e. money in the hand and should be honoured. The trend continues. In *Enco Civil Engineering* v. *Zeus International* (1991) an employer refused to honour a certificate issued by the engineer where the ICE conditions applied. The court did not award summary judgment and referred the matter to arbitration. In the *Bank of East Asia* v. *Scottish Enterprise and Stanley Miller* (1996) it was held that the employer was entitled to deduct the costs of remedying defective work from sums certified.

7.18.6 In the light of the Arbitration Act 1996 courts have no discretion to hear disputes where an arbitration clause exists in the contract. Moreover the decision in *Halki Shipping* v. *Sopex Oils* (1997) leads to the conclusion that if a reason for non-payment is raised the court will be unlikely to investigate whether the reason is bona fide or not. The proceedings will be stayed whilst the matter is referred to arbitration. It would seem that all an unscrupulous employer needs to do is to raise a well orchestrated smoke screen to delay payment.

SUMMARY

In the light of recent case law it would seem that employers may, if a bona fide dispute arises as to the accuracy of a payment certificate, withhold payment whilst the dispute is referred to arbitration.

Moreover the court may be unwilling to investigate whether a reason for non-payment is bona fide or not: *Halki Shipping* v. *Sopex Oils* (1997).

The Arbitration Act 1996 gives comfort to an employer who resists paying a sum certified. Under this Act a court has no discretion to hear a dispute where the contract includes an arbitration clause. The matter must be stayed and referred to arbitration.

The only comfort for the contractor is that it seems now that at the end of the day interest will be awarded from the date the certificate was due for payment (see 7.16).

7.19 Where a cheque is issued in respect of construction work undertaken, can it be stopped before it is honoured if it subsequently becomes obvious that the work is defective?

The case of *Isovel Contracts Ltd* v. *ABB* (2001) involved a dispute concerning payment by cheque for construction work. The work comprised the construction of a leisure centre. ABB was the subcontractor who carried out the mechanical and electrical work, some of which they subcontracted to Isovel. An application for payment was made by Isovel and a cheque in the sum of £70 000 issued by ABB in payment for the work. It appears that under the terms of the subcontract, Isovel was entitled to payment of sums paid to ABB under the main contract by Miller Construction. After ABB had issued the cheque to Isovel, it became clear that Miller had reduced the value of the mechanical and electrical work due to non-compliance. ABB then stopped the cheque made out to Isovel on the basis that the value of the work done by Isovel merited a nil valuation. Isovel commenced proceedings for the payment of the cheque. The judge in the case explained that a cheque is like a bill of exchange and is as good as cash in hand. The party who issued the cheque is not entitled to argue that the work had no value and therefore there is no consideration for the payment, neither can a counter claim be used to avoid payment. ABB was therefore required to make payment of the sum of £70 000 to Isovel.

SUMMARY

Where a cheque is issued it cannot subsequently be stopped if it transpires the work is defective. In *Isovel Contracts Ltd* v. *ABB* (2001) the court considered that a cheque is like a bill of exchange and is as good as cash in the hand.

7.20 Where agreement is reached whereby one party to a construction contract agrees to pay the other a sum of money, can the paying party refuse to make the payment on the grounds that he was financially coerced into the agreement?

7.20.1 If an agreement is reached under duress in the form of economic pressure, it is possible to have the agreement set aside. This could occur where a small contractor has completed the work and is promised payment of a final balance only if he accepts a sum which is less than his due entitlement. As the contractor is short of money, he accepts the offer. Such an agreement is unlikely to be enforceable, leaving the contractor to demand payment of his due entitlement – which was the situation in *D & C Builders Ltd* v. *Rees* (1966).

7.20.2 The case of economic pressure could also be used to secure agreement to a payment in excess of the due entitlement. A subcontractor who has a fixed price contract may coerce the main contractor into agreeing to make additional payments where it was impossible to secure a replacement subcontractor. An agreement reached under these circumstances is unlikely to be enforceable.

7.20.3 The courts will not always hold that an agreement which is secured under financial pressure is unenforceable. In the case of *Williams* v. *Roffey Bros & Nicholls (Contractors) Ltd* (1990) a subcontractor underpriced the carpentry work in connection with the refurbishment of a block of flats. Part way through the work the subcontractor got into financial difficulties and threatened to stop work unless the contractor agreed to increase the contract prices. The contractor considered that, if the subcontractor stopped work, it would cost considerably more to engage a replacement and therefore agreed to an extra payment. The court held that the agreement was binding as the contractor derived the benefit of retaining the subcontractor's services and thus avoided paying more to have the work completed. This seems a strange decision but the judge, who is the only person whose opinion is relevant, considered the facts did not demonstrate economic duress.

7.20.4 It is often a case of the large organisation taking financial advantage of a smaller one, but this is not always the situation.

The case of *Carillion Construction Ltd* v. *Felix (UK)* (2000) tells a familiar tale. Carillion was the main contractor for the construction of an office building for Hammerson UK Properties.

The design, manufacture and supply of the cladding was subcontracted to Felix UK Ltd. Work was commenced by Felix in September 1999, before the subcontract had been entered into. A subcontract became binding upon the parties on 14 January 2000. Complaints were being made by Carillion concerning delays caused by Felix due to late delivery of cladding panels. Disputes were occurring at about this time concerning the likely final value of the work undertaken by Felix. In particular, the parties were experiencing difficulties in reaching agreement on what constituted variations and their value. Much of the work which Felix claimed to be the subject of variations, Carillion considered to be a part of the original works.

Felix, anxious to secure the agreement of Carillion, resorted to what may be considered strong arm tactics. They refused to make deliveries until agreement was reached. Carillion became concerned that there could be a delay to the completion of the main contract works, which could involve payment of liquidated damages to the employer. A significant difference as to the final sum payable to Felix had arisen by the end of March. Carillion considered that a total sum of £2.756m would become payable on completion of the subcontract works. Felix stated that their entitlement was £3.119m. Carillion felt under great pressure to agree a final account.

The parties reached a settlement agreement in the sum of £3.2m which was expressed as full and final. Carillion indicated their displeasure at being forced into the agreement and reverted to their original figures once deliveries had been made. Felix brought an action to try and enforce the agreement. The court, however, considered that the pressure applied by Felix was illegitimate and without justification and refused to enforce the terms of the agreement.

SUMMARY

Courts will not enforce an agreement which has been reached by coercion using financial pressure. However, financial pressure in itself is not sufficient to have the agreement set aside; coercion must also be present.

8
VARIATIONS

8.1 Where a contractor/subcontractor submits a quotation for extra work which is accepted, is the accepted quotation deemed to include for any resultant delay costs?

8.1.1 It is not uncommon, where a major variation is contemplated, for the architect or engineer to call for a quotation from the contractor or subcontractor. Often a lump sum is quoted for work shown on a revised drawing.

If the variation causes delay there is usually an extension of time awarded. The contractor or subcontractor involved sometimes submits a claim for payment of the additional costs associated with the delay. A usual response is that no additional payment is due as the lump sum included or ought to have included for any such additional costs.

8.1.2 Some of the standard forms provide for contractors to quote for variations and cover this point. JCT With Contractor's Design, where Optional Supplementary Provisions S1 to S7 are used, makes it clear in clause 56.3 that the quotation will include the amount of any direct loss and expense which results from the regular progress of the works or any part thereof being materially affected.

The contractor, under clause S6.2.2, has the right to raise an objection to providing a quotation. This seems to recognise that there may well be situations where forecasting the amount of direct loss and/or expense is not possible.

8.1.3 The Engineering and Construction Contract (NEC) in clause 62 provides a detailed procedure for the submission and acceptance of quotations for compensation events and states that such quotations comprise 'proposed changes to the Prices and any delay to the Completion Date assessed by the Contractor'.

The project manager has to reply within two weeks of the submission and will either:

- instruct the contractor to submit a revised quotation,
- accept the quotation,
- give notice that the proposed instruction or proposed changed decision will not be given, or
- give notice that the project manager will make his own assessment.

The project manager may extend the time for submission of the quotation in his reply.

8.1.4　Where the conditions are silent on the matter a contractor or subcontractor could refuse to submit a quotation.

Where quotations are requested for extra or varied work it may be argued that a separate contract is contemplated. This being the case, in the absence of any wording to the contrary, if the quotation is accepted the contractor/subcontractor will be required to do everything necessary to carry out the extra work. If the extra or varied work disrupts or delays the contract work it seems reasonable that the quotation should include for any resultant costs which are foreseeable. Over and above such additional costs the contractor/subcontractor, if this argument is sustainable, will be entitled to payment of additional unforeseen costs.

Courts, however, seem nowadays to be endeavouring to resolve disputes within the context of the contract. A more likely interpretation is that the quotation and acceptance represents agreement of a financial entitlement which the contractor/subcontractor has under a clause or clauses in the contract.

8.1.5　In the case of the ICE 6th Edition a logical argument is that the agreement relates to the evaluation of a variation under clause 52.1. However, it seems unlikely that a quotation would be held to include any adjustment of rates which the engineer has power to fix where the varied works renders inappropriate or inapplicable other parts of the work.

The ICE 7th Edition, under clause 53, empowers the engineer to request the contractor to submit a quotation for a proposed variation and the contractor is specifically required to include in his quotation an estimate of any delay and the cost of such delay.

8.1.6　Where JCT 80 applies it would seem reasonable to interpret an agreement as being intended to cover the evaluation of the variation under clause 13.5.1. It would be unreasonable, in the absence of express wording, for the contractor to be able to foresee the extent to which the variation would substantially change the conditions under which any other work is executed and therefore how much additional payment may be due under clause 13.5.5.

Difficulty may also be experienced in foreseeing the amount of direct loss and/or expense which may arise due to the regular progress of the work being materially affected by a variation. In the absence of express wording it is unlikely that a quotation would be held to include direct loss and/or expense under clause 26.

8.1.7　JCT 98 also now provides for the contractor to quote for variations. Clause 13A.2.2 states that a quotation must include any adjustment to time and clause 13A2.3 provides for an amount to be paid in respect of loss and expense caused by delays to the progress of the work.

SUMMARY

It may be argued that, in the absence of a clause in the contract which caters for quotations being submitted, such an agreement constitutes a separate contract. This being the case, the contractor/subcontractor would be required to include for everything necessary to carry out the extra or varied work. In all probability it would be expected that foreseeable delay to the contract works would be catered for in the quotation. Any unforeseeable additional costs would be the subject of a separate claim.

A more plausible argument is that the quotation is in respect of an entitlement which the contractor/subcontractor has under one or more of the clauses in the contract. In the case of the ICE 6th Edition the quotation may be with regard to the valuation of a variation under clause 52.1. It would seem, in the absence of wording to the contrary, unlikely to include any adjustment of rates which the engineer has power to fix where the varied work renders inappropriate or inapplicable other parts of the work. Where JCT 80 applies an agreed quotation would seem to include for a valuation under clause 13.5.1. In the absence of express wording it is unlikely to be held to include additional payment for disruption under clause 13.5.5 or direct loss and/or expense under clause 26 due to the difficulty in most cases of foreseeing the likely effects and resultant additional costs with any sort of accuracy.

The recent revisions, JCT 98 clause 13A and the ICE 7th Edition, provide for quotations in respect of variations and such quotations must include for time delays and additional cost.

Contractors and subcontractors, when submitting quotations for extra works, should make it expressly clear to what extent the quotation includes for loss and expense resulting from delay and disruption.

8.2 Can a contractor/subcontractor be forced to carry out a variation after practical completion?

8.2.1 Once practical completion has been achieved it is convenient to have the contractor available to carry out extra work during the defects period. This may be for example to correct a design fault or to comply with the occupier's requirements. In many instances this may be inconvenient for the contractor or they could consider that they have been ill-served by the employer's quantity surveyor and so have no wish to do any more work. On the other hand the contractor is familiar with the project and would normally be the most suitable organisation to carry out the work.

8.2.2 Once practical completion has been achieved, however, the contractor has no obligation to carry out varied work instructed by the architect/engineer. An exception would be where the contract expressly provides for variations being issued post practical completion.

8.2.3 *Hudson's Building and Engineering Contracts* 11th edition at paragraph 4.182 states:

> 'The last of the above mentioned cases [*SJ and MM Price Ltd* v. *Milner* (1968)] though somewhat inadequately reported supports the view that variations as well as original contract work cannot be instructed after practical completion of the remainder of the work in the absence of express provision, unless of course the contractor is willing to carry them out.'

SUMMARY

The contractor is not obliged to carry out variations where the instruction is issued after practical completion unless there is a clause in the contract which gives the architect/engineer power to issue an instruction of this nature.

8.3 Where work is omitted from the contract by way of a variation can a contractor/subcontractor claim for loss of profit?

8.3.1 Contractors often argue that where work is omitted from their contract they lose an opportunity of earning the profit element which was built into the value of work omitted. This being the case they claim from the employer the loss they allege to have been suffered.

8.3.2 Whether or not the contractor is entitled to the loss of profit is not clear cut.
ICE 7th Edition deals with the evaluation of variations in the following clauses:

- 52(3) – new work to be valued at rates and prices set out in the contract
- 52(4) – the engineer has power to change rates or prices for any item of work in the contract which are rendered inapplicable due to the varied work.

The equivalent clauses in ICE 6th Edition are 52(1) and (2).
These clauses are not too helpful in answering the question as to whether loss of profit should be paid where work is omitted.
In the case of *Mitsui Construction Co Ltd* v. *The Attorney General of Hong Kong* (1986), the court seemed to give the engineer wide scope when exercising his powers under the equivalent of clause 52(4) of the ICE conditions to adjust contract rates. A reasonable argument may be that such adjustment should be made to take account of lost profit.

8.3.3 Where JCT 98 applies, clause 13 deals with variations and stipulates that they will be valued at bill rates where work is of a similar character, executed under similar conditions and does not significantly change the quantities. If there is a significant change of quantities then the contractor may become entitled under clause 13.5.1.2 to a variation to the rate to include a fair allowance for the change of quantities. It may be argued that the fair valuation should include the loss of profit in respect of work omitted.
Clause 13.5.5 provides for adjusting contract rates where the conditions have been substantially changed due to a variation. The wording differs from the clauses in the ICE conditions and is not therefore helpful to contractors wishing to argue that contract rates should be amended to take account of lost profit arising from omitted work.

8.3.4 In the case of *Wraight Ltd* v. *PH & T Holdings* (1968) a contractor's contract was wrongly determined with the work part completed. The determinations clause provided for the contractor to be paid 'any direct loss and/or damage caused to the contractor by the determination'. It was held by the court that this wording included loss of gross profit on the uncompleted work.

8.3.5   JCT 63 under clause 11(6) allows the contractor to recover direct loss and/or expense arising from a variation. Following the *Wraight* case this would include loss of gross profit. Therefore if the contractor could show that as a result of an omission profit had been lost the loss could be recovered if the contract were worded in a like manner to JCT 63.

JCT 98, however, does not include a clause equivalent to clause 11(6) of JCT 63. Clause 26 of JCT 98 deals with loss and expense resulting from a variation but only applies where the regular progress of the works has been materially affected.

8.3.6   On loss of profit in a wider context, in *Bonnells Electrical Contractors* v. *London Underground* (1995) it was held that where a call out contract was wrongly determined the injured party was entitled to loss of profit on work which would have been carried out during the period of notice which ought to have been given.

SUMMARY

The answer to the question is not straightforward but depends upon the wording of the contract. Standard forms in general use provide for the work to be varied including omissions and therefore there is no scope for claiming damages for breach.

Contracts which are worded along the lines of JCT 63 clause 11(6) allow the contractor to claim loss and expense where the work is varied. In *Wraight Ltd* v. *PH & T Holdings* (1968) it was held that the wording 'direct loss and/or damage' included gross profit. Therefore it would seem that clause 11(6) of JCT 63 would also allow for loss of profit.

Where wording akin to clause 11(6) is not included in the contract the net has to be cast wider by contractors looking for friendly wording in the contract. For example the ICE 6th and 7th Editions, by clause 52, provide for adjusting contract rates rendered inappropriate due to the varied work. Contractors may argue that in adjusting rates allowance should be made for loss of profit from work omitted. Under JCT 98, by clause 13.5.1.2, the contractor would have an entitlement to a fair valuation where a significant omission occurred. This, it may be said, should cater for lost profit from work omitted.

It is, moreover, quite common where substantial work is omitted for engineers and quantity surveyors to make allowance for loss of profit.

8.4   Where work is given to another contractor is there a liability to pay loss of profit?

8.4.1   It is within the powers of the parties to enter into a contract whose terms give the employer the right to omit work and have it carried out by others. If this be the case the terms of the contract should then go on to indicate whether or not the contractor is entitled to claim loss of profit. The standard forms of contract in current use do not include a provision of this nature. In the absence of such a clause what remedy, if any, do contractors have where work included in their contract is omitted and the employer arranges to have it carried out by others?

8.4.2 This matter was one of the subjects of dispute in *Amec Building Ltd* v. *Cadmus Investments Co Ltd* (1996) which was referred to arbitration.

The arbitrator awarded Amec sums for loss of profit in connection with a food court. The work was covered by certain provisional sums, but an architect's instruction of 9 April 1990 omitted the work from the contract. This was subsequently let to another contractor, and Amec successfully claimed loss of profit of over £12 800, plus statutory interest. There had been no agreement between Amec and Cadmus that the work should be omitted.

In finding in favour of Amec on appeal the court held:

> There is no dispute that the power is given to the architect in his sole discretion to withdraw any work from provisional sums for whatever reason if he considers it in the best interests of the contract or the employer to do so. The difficulty that arises in this case is that which arose in the Australian case [*Carr* v. *JA Berriman Pty Ltd* (1953)], namely that it would appear that the purpose was to remove it from the existing contractor and award the work to a new contractor. Without a finding that the architect was entitled to withdraw the work for reasons put forward by [counsel], and in view of the fact that the specific reasons he advances were expressly rejected by the arbitrator, it seems to me that the only conclusion I can come to is that the arbitrator had concluded that it was an arbitrary withdrawal of the work by the architect in order to give it to a third party other than Amec. In those circumstances, and, in particular, in view of the express finding of the arbitrator at paragraph 12.04 that the statement in Hudson reflects the 'generally accepted position in the industry', it seems to me that the arbitrator was perfectly correct in deciding that such an arbitrary withdrawal of work from the provisional sums and the giving of it to the third party was something for which Amec were entitled to be compensated and the compensation that he arrived at, namely the loss of the profit having accepted figures put forward to him in evidence, is one which is not open to be impugned on appeal as a matter of law. In those circumstances, therefore, albeit with some reluctance, it seems to me that I should dismiss the appeal as well.

8.4.3 It seems clear from this decision and the Australian case of *Carr* v. *JA Berriman Pty Ltd* (1953) to which the judge made reference that if work is omitted from the contract and given to others then the contractor will be entitled to claim loss of profit.

SUMMARY

The parties are at liberty to enter into a contract which allows the employer to omit work included in the contract and arrange for it to be carried out by others. Such a clause should state whether or not the contractor is entitled to claim for loss of profit. The standard forms in general use do not include such a clause. It has been held in a legal case that where work is omitted and given to others the contractor is entitled to claim for loss of profit.

This also applies to provisional sums omitted with the work given to others.

8.5   Where due to a variation a contractor has to cancel an order for the supply of material can he pass on to the employer a claim received from the supplier for loss of profit?

8.5.1   Variations issued by architects or engineers often put contractors to substantial additional expense. Standard forms of contract are not always precise as to how variations are to be valued. Often arguments take place as to whether a particular item of expenditure may be recovered.

8.5.2   The matter was considered by the Court of Appeal when it had to make a decision on a disputed item of cost resulting from a variation in the case of *Tinghamgrange Ltd (T/A Gryphonn Concrete Ltd)* v. *Dew Group and North West Water* (1996). As a result of a variation the contractor had to cancel an order from a supplier who claimed for loss of profit. The contractor sought reimbursement from the employer and joined them as third party to the action.

8.5.3   Early in 1989 North West Water commissioned works at Oswestry Water Treatment Works. Dew were the main contractors and the contract incorporated the ICE 5th Edition.

Part of the contract works involved new pre-cast concrete under drainage blocks which had to be specially manufactured. North West Water inspected Gryphonn's premises and Dew placed an order with Gryphonn worth over £250 000.

The reverse of Dew's order stated:

> '14. In the event of contract works in connection with which this order is given being suspended or abandoned, the contractor shall have the right to cancel any order by written notice given to the subcontractor, vendor or supplier, on any such cancellation, the subcontractor, vendor or supplier shall be entitled to payment as provided by the terms of the main contract. No allowance, however, will be made on account of loss of profit on uncompleted portions of the order.'

Gryphonn had manufactured about one quarter of the blocks, most of which were either delivered to site or were about to be delivered when North West Water's resident engineer instructed Dew to cancel the order with Gryphonn as the specification had been changed. Dew cancelled the order and wrote to North West Water stating that the change in the specification could lead to additional costs.

Gryphonn claimed about £50 000 for loss of profit on the blocks which had been ordered but not delivered, plus £1740 for the cost of the mould manufactured to satisfy Dew's order. Dew passed this claim to North West Water. North West Water eventually paid for the cost of the mould, but refused to make any payment in respect of loss of profit emanating from the cancellation of the order.

8.5.4   The matter went before the County Court in Newport where Gryphonn claimed its loss of profit from Dew. Dew relied upon clause 14 of the contract and sought an indemnity from North West Water.

Gryphonn was awarded nearly £40 000, plus interest, but the court declined to make any award by way of indemnity in the third party proceedings. Whilst the judge expressed sympathy with Dew, as it had merely done as instructed by the engineer, there was no main contract provision which enabled the court to make any award in respect of Gryphonn's loss of profit claim.

8.5.5 Dew appealed, arguing that its claim was simply for the cost of the work it had done in accordance with the instructions received. Paying Gryphonn's loss of profit was an integral part of that cost and to exclude that part from the valuation of the work was unfair.

It was held that:

> Clause 51 of the contract . . . between North West Water Ltd. . . . and the Dew Group Ltd gives the employer the right to make any variation comprised within the wide scope of clause 51(1). It is unnecessary to decide in this appeal whether a contract containing such a clause would have been legally binding – one party having the right unilaterally to vary any term of the contract – if it had not been for its inclusion of a further provision, expressed in clause 52(1), whereby a 'fair valuation' has to be made. The 'fair valuation' of the variation is intended to provide a fair compensation to the contractor for any adverse financial effect upon it, resulting from the unilateral variation. We assume, therefore, . . . that the validity of the head contract is not affected by clause 51, because of the provision for fair compensation in the event of loss resulting to the contractor from a unilateral variation.
>
> 'The employer was, or should have been, well aware that the contractor would not itself be the supplier of the concrete blocks, but that these would be purchased under a subcontract. The employer was, or should have been, equally aware that the result of the variation which was made – the cancellation of the order for 209 874 concrete blocks – would necessarily result in a loss of profit for the subcontractor, Gryphonn, since the contractor would be contractually bound under the head contract to pass on to the subcontractor the variation consisting of the cancellation.'

The court found in favour of Dew who were entitled to recover from North West Water the sum paid to their supplier Gryphonn for loss of profit arising from the cancellation of part of the order for supplying blocks.

SUMMARY

The engineer when carrying out a fair valuation under the ICE conditions in respect of a variation to omit work which includes materials received from a supplier should provide fair compensation to the contractor for any adverse financial effect upon it resulting from the variation. This would include a legitimate loss of profit claim from the supplier. There is no reason why this principle should not apply to JCT and GC Works/1 conditions.

8.6   How are 'fair' rates defined?

8.6.1   Most forms of contract provide for the contract documents to include contract rates which are to be applied when evaluating variations. These
contract rates are usually employed where work in the variation is similar
to work set out in the contract. For example JCT 98 clause 13.5.1.1 requires
the work to be of a similar character and carried out under similar conditions to work described in the contract documents.

Work included in a variation which differs in character to work in the
contract is usually expressed as being evaluated using fair rates e.g. JCT 98
clause 13.5.1.3.

8.6.2   Disputes often arise where fair rates are to be employed as to how the fair
rate will be calculated. Courts, if asked to decide on the matter, would
invariably take a subjective view and hold that the rate must be fair in all the
circumstances which occurred on the project and which were relevant to the
calculation of the rate.

8.6.3   In the case of *Semco Salvage & Marine Pte* v. *Lancer Navigation Co Ltd* (1997) the
House of Lords had to decide the meaning of fair rates. A dispute arose in
connection with a salvaging operation at sea. The parties could not agree to
the amount of reimbursement to a company undertaking salvaging operations. How a 'fair rate' as referred to in the Lloyds Open Form 1990 was to be
calculated formed the basis of the disagreement.

Clause 1(a) of the Form states:

'... The services shall be rendered and accepted as salvage services upon the
principle of 'no cure–no pay' except where the property being salved is a tanker ...
[when] ... the contractor shall, nevertheless, be awarded solely against the owners
of such tanker his reasonably incurred expenses and an increment not exceeding
15% of such expenses ... Within the meaning of the said exception of principle of
'no cure–no pay', expenses shall in addition to actual out of pocket expenses
include a fair rate for all tugs, craft, personnel and other equipment used by the
contractor in the services ...'

The principal issue before the court was the definition of 'expenses' in article
14.3 of the International Convention on Salvage, particularly the part of it
which includes in the expenses 'a fair rate for equipment and personnel
actually and reasonably used in the salvage operation'.

8.6.4   Whilst the judgment was given with particular reference to the wording of
the Convention and in relation to salvage, the House of Lords made some
general observations as to the meaning of 'fair rate'.

Lord Lloyd of Berwick had this to say:

(1)   ... 'fair rate for equipment and personnel actually and reasonably used in the
salvage operation' in article 14.3 means a fair rate of expenditure, and does
not include any element of profit. This is clear from the context, and, in
particular, from the reference to 'expenses' in article 14.1 and 2, and the
definition of 'salvors' expenses' in article 14.3. No doubt expenses could have
been defined so as to include an element of profit, if very clear language to
that effect had been used. But it was not. The profit element is confined to the
mark-up under article 14.2, if damage to the environment is minimised or
prevented.

(2) The first half of article 14.3 covers out-of-pocket expenses. One would expect to find that the second half of the paragraph covered overhead expense. This is what it does. If confirmation is needed, it is to be found in the reference to sub-paragraphs (h) to (j) of article 13.1 . . .

(3) [Counsel for the salvage operators] argued that the word 'rate' included more naturally a rate of remuneration rather than a rate of expenditure. But, as Lord Mustill points out, 'rate' is the appropriate word when attributing or apportioning general overheads to the equipment and personnel actually and reasonably used on the particular salvage operation.

(4) [Counsel] argued that, if a fair rate means 'rate of expenditure', it would require 'a team of accountants' in every salvage arbitration where the environment has been at risk. [Opposing Counsel's] answer was that the basic rates in the present case . . . were agreed without difficulty between the two firms of solicitors. In any event, accountants are nowadays, as he says, part of everyday life.

8.6.5 In the *Semco Salvage* case the court clearly defined the words 'fair rate' in the context of the contract. The exclusion of profit should not, therefore, be taken as of general application when defining fair rate under different circumstances. In deciding what was a fair rate, the amount of renumeration was, in the end, preferred to the amount of expenditure, but profit was excluded.

8.6.6 By way of contrast in the case of *Tinghamgrange Ltd* v. *Dew Group and North West Water* (1996) the Court of Appeal held that in the ICE 5th Edition a 'fair valuation' under clause 52(1) included compensation to the main contractor for a loss of profit payment to a subcontractor in respect of the cancellation of an order resulting from an engineer's variation. (See also 8.5 in respect of a supplier's loss of profit claim.)

8.6.7 In the case of *Banque Paribas* v. *Venaglass Ltd* (1994) the court had to decide how the 'fair and reasonable value' should be calculated in respect of a part-completed project. It was necessary to carry out the valuation as the developer had become insolvent. In the absence of any guidelines in the development agreement the judge decided that a fair and reasonable value should be calculated on a cost or measure and value basis and not on an open market value of the project.

8.6.8 Judge Bowsher QC in his judgment in the case of *Laserbore Ltd* v. *Morrison Biggs Wall Ltd* (1992) had to decide the meaning of 'fair and reasonable payments for all works executed'.

He had this to say regarding costs as being the correct method of fixing a fair and reasonable payment:

> I am in no doubt that the costs plus basis in the form in which it was applied by the defendant's quantum expert (though perhaps not in other forms) is wrong in principle even though in some instances it may produce the right result. One can test it by examples. If a company's directors are sufficiently canny to buy materials from stock at knockdown prices from a liquidator, must they pass on the benefit of their canniness to their customers? If a contractor provides two cranes of equal capacity and equal efficiency to do an equal amount of work, should one be charged at a lower rate than the other because one crane is only one year old but the other is three years old. If an expensive item of equipment has been depreciated to nothing in the company's accounts but by careful maintenance the company continues to use it, must the equipment be provided free of charge apart from the

running expenses (fuel and labour)? On the defendant's argument the answer to those questions is 'yes'. I cannot accept that that begins to be right.

8.6.9   It can be confusing for two judges to offer differing opinions concerning a legal interpretation of commonly used wording. In the *Laserbore* case, Judge Bowsher was not in favour of applying a cost reimbursable method in respect of 'fair and reasonable payments'.

A differing approach was taken in *Weldon Plant Ltd* v. *The Commissioner For The New Towns* (2000). In this case a dispute arose in connection with the valuation of a site instruction to excavate gravel and backfill with clay in connection with the construction of the Duston Mill Reservoir. The contractor was entitled to be paid on a fair valuation basis.

The judge took the view that, in evaluating a fair valuation, the calculation should be based upon the reasonable costs of carrying out the work if reasonably and properly incurred. If, in the execution of the work, cost or expenditure is incurred which would not have been incurred by a reasonably competent contractor in the same or similar circumstances, then such costs would not form part of a fair valuation. Attention was drawn to what *Keating on Building Contracts* has to say on the constituent elements of a fair valuation. It states that useful evidence may include a calculation based on the net cost of labour and materials used, plus a sum for overheads and profit. The judge considered that a fair valuation should include elements for the cost of labour, the cost of plant, cost of materials, the costs of overheads and profit. In his judgement, a fair valuation has not only to include something on account of each of those elements, but also it would not be a fair valuation within the meaning of the contract if it did not do so.

8.6.10   A different argument was preferred in the case of *Seck Controls Ltd* v. *Drake and Scull Engineering Ltd* (2000). It was held that Seck was entitled to be paid a reasonable sum for the work it had completed. Drake and Scull was of the opinion that payment should be based upon normal and usual rates.

This was not accepted by the judge. In assessing Seck's entitlement he commenced in many instances with Seck's costs. However, there were many allegations of inefficiency and wasted cost and the judge had to consider whether there was any precedence in English law for making adjustments to take account of such matters. The law he considered was undecided as explained in the decision in *Crown House* v. *Amec Projects* (1989). He was, however, influenced by the decision in *Lachhani* v. *Destination Canada* (1996). It was held in this case that where it was appropriate to calculate a fair valuation from costs actually incurred, then they would have to be reasonably, necessarily and properly incurred. If the building contractor works inefficiently and/or if the building contractor leaves defective work, then quite obviously the costs incurred must be appropriately adjusted as the owner should not be expected to pay more than the construction work is worth.

SUMMARY

The House of Lords in a salvage case held that the words 'fair rates and expenses' mean a fair rate of remuneration, not expenditure, but does not include loss of profit. The wording could have been defined in the contract to include profit but this was not the case.

The judge in the *Banque Paribas* case preferred to employ a measure and value method to establish the value of a development and not an open market value basis but in *Laserbore* the judge was not impressed by the argument that fair and reasonable payments should be based upon costs.

In *Weldon Plant* it was held that a fair valuation should be based upon the reasonable costs of carrying out the work if reasonably and properly incurred. If the contractor works inefficiently or carries out work defectively, then in accordance with the *Seck Controls* decision, the costs should be adjusted.

These cases cannot be taken as definitions to be used universally. Courts will normally take a subjective view and hold that the rate must be fair in all the circumstances which occurred on the particular project and which were relevant to the calculation of the rate.

8.7 When do *quantum meruit* claims arise and how should they be evaluated?

8.7.1   The expression '*quantum meruit*' means 'the amount he deserves' or 'what the job is worth' and in most instances denotes a claim for a reasonable sum.

Payment on a quantum meruit basis will normally arise in circumstances where a benefit has been conferred which justice requires should result in reimbursement being made. It does not usually arise if there is an existing contract between the parties to pay an agreed sum.

There may, however, be a quantum meruit claim in the following circumstances:

- **An express agreement to pay a 'reasonable sum'.**
- **No price fixed**. If the contractor does work under a contract, express or implied, and no price is fixed by the contract, he is entitled to be paid a reasonable sum for his labour and the materials supplied.
- **A quasi-contract**. This may occur where, for instance, there are failed negotiations. If work is carried out while negotiations as to the terms of the contract are proceeding but agreement is never reached upon essential terms, the contractor is entitled to be paid a reasonable sum for the work carried out: *British Steel* v. *Cleveland Bridge* (1984), *Kitson Insulation Contractors Ltd* v. *Balfour Beatty Ltd* (1990), *Monk Construction Ltd* v. *Norwich Union* (1992).
- **Work outside a contract**. Where there is a contract for specified work but the contractor does work outside the contract at the employer's request the contractor is entitled to be paid a reasonable sum for the work outside the contract on the basis of an implied contract. In *Parkinson* v. *Commissioners of*

Works (1949) the contractor agreed under a varied contract to carry out certain work to be ordered by the Commissioners on a cost plus profit basis subject to a limitation as to the total amount of profit. The Commissioners ordered work to a total value of £6 600 000 but it was held that on its true construction the varied contract only gave the Commissioners authority to order work to the value of £5 000 000. It was held that the work that had been executed by the contractors included more than was covered, on its true construction, by the variation deed, and that the cost of the uncovenanted addition had therefore to be paid for by a *quantum meruit*.

8.7.2   In the case of *Laserbore Ltd* v. *Morrison Biggs Wall Ltd* (1992) a subcontractor in accordance with a letter of intent was entitled to be paid 'fair and reasonable payments for all works executed'.

The contract referred to in the letter of intent was to be based upon the FCEC Blue Form but no price or mechanism for agreeing the price was reached. It was the contractor's argument that the subcontractor's recorded costs should be the basis of payment. Judge Bowsher QC did not agree for the reasons as detailed in 8.6.8.

8.7.3   In the decision of *Weldon Plant Ltd* v. *The Commissioner For The New Towns* (2000) it was held that a fair and reasonable payment should be based upon the reasonable costs of carrying out the work if reasonably and properly incurred. The case of *Seck Controls Ltd* v. *Drake & Scull Engineering Ltd* (2000) again addressed the matter of the meaning of reasonable costs. It was held that, where it was appropriate to calculate a fair valuation from costs actually incurred, they would have to be reasonably, necessarily and properly incurred. If the contractor worked inefficiently or constructed work defectively, then an appropriate adjustment must be made.

8.7.4   A 'contractual quantum meruit' was referred to in *ACT Construction Ltd* v. *E. Clarke & Son (Coaches) Ltd* (2002) where there was no formal contract. The Court of Appeal held that, provided there is an instruction to do work and an acceptance of that instruction, then there is a contract, and the law will imply into it an obligation to pay a reasonable sum for the work undertaken. It was decided that reasonable remuneration involved the sum of the contractor's costs plus 15%.

Another Court of Appeal case, that of *A.L. Barnes Ltd* v. *Time Talk UK* (2003), related to work being undertaken without a contract. In this case, account had to be taken of an overpayment in arriving at a reasonable remuneration.

8.7.5   Contractors having entered into a contract with all the terms including the price agreed, have been known to argue that due to changes of a critical nature and serious delays caused by the employer the contract has been frustrated. If the argument holds water the contract rates will no longer apply and payment should be made on a quantum meruit basis. This argument was the basis of the case of *McAlpine-Humberoak* v. *McDermott International* (1992) and which succeeded in the lower court. The Court of Appeal did not reject the principle of payment on a quantum meruit basis but found on the facts that no frustration had occurred.

8.7.6   Where work has been carried out and there occurs a breach giving rise to a discharge of the contract, the injured party may claim for payment on a quantum meruit basis as an alternative to a claim for loss of profit on the uncompleted work. This can be useful to a contractor or subcontractor where they have carried out part of the work on uneconomical rates and the other party then repudiates the contract resulting in determination. Payment for the work carried out up to the repudiation may be made on a quantum meruit basis. The contract rates in this case are no longer applicable though they may be used as evidence of what should be a reasonable remuneration for the work.

8.7.7 On the other hand in the case of *Lachhani* v. *Destination Canada UK* (1996) negotiations for construction work broke down and no contract was concluded. Nonetheless the work was carried out and payment was due on a quantum meruit basis. The contractor's claim for payment was well in excess of the amount of its quotation. The court held that a building contractor should not be better off as a result of a failure to conclude a contract than he would have been if his quotation been accepted. It was the court's view that the amount of the quotation must be the upper limit of his entitlement. In some cases it may be appropriate to consider the level of pricing being negotiated before work is started.

8.7.8 When evaluating a quantum meruit claim the question may arise as to whether it should be based upon the value to the recipient or the cost to the party doing the work. The Australian case of *Minister of Public Works* v. *Lenard Construction* (1992) opted for the value to the recipient.

8.7.9 The following principles concerning quantum meruit were established in another Australian case, *Brenner* v. *Firm Artists Management* (1993):

(1) A claim for quantum meruit presupposes that no contract exists.

(2) The yardstick for determining a claim for quantum meruit is what is a fair and reasonable remuneration or compensation for the benefit accepted, actually or constructively.

(3) If the parties have agreed a price for certain services and the services are performed the agreed price may be given as evidence of the appropriate remuneration but is not in itself conclusive.

(4) The appropriate method of assessing benefit in some cases may be by applying an hourly rate to the time involved in performing the services. Where it is difficult to determine the number of hours involved the court may make a global assessment.

SUMMARY

There is no hard and fast rule as to the basis upon which a quantum meruit valuation is to be based. The court will take into account all the circumstances. It may be that a contractor's costs are a fair and reasonable method of evaluating a quantum meruit valuation but not by any means the only method. The court may decide that the value of the project to the recipient of the work is to be preferred to the contractor's costs as the basis for fixing a quantum meruit payment. Where a quotation has been submitted but no contract concluded the sum fixed for a quantum meruit payment is unlikely to be greater than the quotation.

8.8   Can the issue of a variation to the work ever have the effect of creating a separate or replacement contract?

8.8.1   Most standard forms of contract include a clause under which the employer or his representative is able to issue an instruction to the contractor to vary the works which are described in the contract. A change in shape of the scheme, the introduction of different materials, and revised timing and sequence are all usually provided for by the variations clause. It will also usually include a mechanism for evaluating the financial effect of the variation and there is normally provision for adjusting the completion date. In the absence of such a clause the employer could be in a difficulty should a variation to the works be required. The contractor could either refuse to carry out the work or undertake the work and insist upon payment on a quantum meruit or fair valuation basis. Calculation of the price for the extra work applying this method could involve payment well in excess of the contract rates.

8.8.2   Even where a contract includes the usual variations clause there may be circumstances which could lead to additions or changes introduced by the employer which fall outside the variations clause. Contractors who find themselves with unattractive contract prices would find it to their advantage to argue that a change introduced by the employer fell outside the variations clause, thus leaving the way open to argue that payment for the change should be on a quantum meruit or fair valuation basis.

8.8.3   In the case of *Thorn* v. *Mayor & Commonalty of London* (1876) the judge said, 'If the additional or varied work is so peculiar, so unexpected, and so different from what any person reckoned or calculated on, it may not be within the contract at all'. A mere increase in the quantities of the work will not invalidate the original contract, even though substantial. The work ordered must be totally different from that contracted for. This being the case, the contractor will be entitled to payment on a quantum meruit basis and not the contract rates.

8.8.4   In *Bush* v. *Whitehaven Trustees* (1888), a contractor undertook to lay a conduit pipe in the month of June and, because of delays on the part of the employer, was unable to proceed before the winter when wages were higher and the works more difficult due to weather conditions. The court held that these changed circumstances were such that the contractor became entitled to payment on a quantum meruit basis.

8.8.5   In the case of *Blue Circle* v. *Holland Dredging Co* (1987) the contractor was required to remove large quantities of excavated material from the site as part of the contract. A variation was issued by the engineer to the effect that the excavated material should be used to form a bird island instead of removing it from the site. Again, the court held that this variation amounted to a separate contract.

8.8.6   By way of contrast, in the case of *McAlpine Humberoak Ltd* v. *McDermott*

International (1992) a contract was let for the construction of four huge pallets in connection with the construction of an oil rig. The number of drawings used increased from 22 to 161, which led to many technical queries. Further, the number of pallets was reduced from four to two. The lower court was receptive to the idea of a separate contract, but this was overruled by the Court of Appeal.

8.8.7   The position in the USA is more developed. It addresses a situation where a large number of changes are instructed which individually fall within the ambit of the variations clause but collectively have the effect of completely changing the scope of the works. This situation is referred to as either abandonment or cardinal change and deals with the situation where the employer makes excessive changes to a project beyond what the parties reasonably could have anticipated at the time the contract was entered into. Courts will look at a number of factors in helping to decide whether the changes have been excessive. The starting points are the size, complexity and expected duration of the contract. Other factors to be considered are the number of changes, how many changes were anticipated when the project started, the magnitude of the work involved in the changes and the length of time over which such changes were made. There is no required intention on the part of the employer to abandon the contract by introducing excessive changes; this will often be implied as a result of constant interference or change. If the parties ignore the procedural provisions of the contract with regard to variations this could help influence the court into accepting that abandonment has occurred.

SUMMARY

It is difficult to be hard and fast as to when additional or changed work will constitute a separate contract or convert the contract into a different one. Courts in the USA seem to be more sympathetic to the contractor's case for abandonment or critical change than in most other countries. Courts and arbitrators in the UK and like jurisdictions find themselves in the long grass when trying to decide what is due if payment is to be on a quantum meruit or fair valuation basis. They feel more comfortable in dealing with additions or changes priced at contract rates and are inclined to play it safe in holding that the facts as presented have not resulted in a separate contract. Employers who are perhaps starting to feel uneasy about the prospect of the USA attitudes creeping into the thinking of judges or arbitrators could give consideration to rewording the variations clause to give a wider definition of additions and changes.

9

LOSS AND EXPENSE

9.1 Where a contractor/subcontractor is granted an extension of time, is there an automatic right to the recovery of loss and expense?

9.1.1 There is a common misconception in the construction industry that once an extension of time has been granted, there will be an automatic entitlement to the recovery of loss and expense. Under most standard forms of contract, the entitlements to extensions of time and payment of additional cost or loss and expense are quite separate. Most of the JCT forms of contract provide for extensions of time under clause 23, with loss and expense dealt with under clause 25.

9.1.2 Many of the reasons for awarding extensions of time also provide an entitlement to additional cost or loss and expense, an example being the late issue of drawings by the architect or engineer. Some grounds for extension of time, for example *force majeure* and excessively adverse weather under JCT contracts, give an entitlement to more time but no facility for financial recovery.

9.1.3 The Society of Construction Law Delay and Disruption Protocol 2002 makes the situation clear when it states:

> '1.6.2   Entitlement to an EOT does not automatically lead to entitlement to compensation (and vice versa).'

9.1.4 The Engineering and Construction Contract is an exception. Within each of the alternative main contract and subcontract options there is a compensation clause. Where one of the events listed in the compensation clause occurs, the contractor/subcontractor is entitled to the resultant extra time and money.

SUMMARY

A contractor/subcontractor is not, under most standard forms of contract, entitled to the recovery of loss and expense. The Engineering and Construction Contract is an exception where the compensation clause provides for both extensions of time and the recovery of additional cost.

9.2 Where a contractor/subcontractor successfully levies a claim against an employer for late issue of drawings, can the sum paid out be recovered by the employer from a defaulting architect/engineer?

9.2.1 Employers who find themselves having to make payments to contractors as a result of the late issue of drawings by the architect or engineer usually feel aggrieved. They often contemplate sending a claim for payment to the architect or engineer or plan to deduct such claims from fees as they fall due. For the employer to have a legal right to take such action he must be able to show that the late issue of drawings by the architect or engineer was a breach of duty.

9.2.2 The duties owed by architects, engineers and other designers to their clients are either express or implied in their terms of engagement.

The Appointment of an Architect SFA/99 published by the RIBA provides in condition 2.1:

'The Architect shall in performing the services ... exercise reasonable skill and care in conformity with the normal standards of the Architect's profession.'

9.2.3 Late issue of drawings may result from a number of causes other than breach of duty by the architect/engineer.

- The employer may have delayed making a decision or introduced a late change.
- There may have been a change introduced by statute or the fire officer.
- Information from the consulting engineer or statutory authority may have caused the delay.

Late issue of drawings does not automatically mean that the architect/engineer is guilty of breach of duty.

9.2.4 Employers when appointing an architect/engineer may wish to be a little more precise in setting out their duties than the general wording provided in the RIBA Appointment of an Architect form.

JCT 98 edition provides in the sixth recital:

'the Employer has provided the Contractor with a schedule ("Information Release Schedule") which states what information the Architect will release and the time of that release.'

This clause is optional but if it is used it may be in the employer's interests to write a clause into the architect's conditions of engagement providing an obligation to issue drawings to conform with the information release schedule.

9.2.5 Where wording of a general nature is included in the conditions of appointment such as the RIBA wording requiring the architect to exercise reasonable skill and care in the normal standards of the architect's profession, it will be necessary for an employer, if he is to be successful, to show that the drawing production by the architect fell short of what one could expect from the ordinary skilled architect.

9.2.6   In the case of *London Underground* v. *Kenchington Ford* (1998) Kenchington Ford were appointed to provide civil engineering and architectural design services in connection with the Jubilee Line station at Canning Town. They were under an express duty set out in their terms of engagement to exercise all reasonable professional skill and diligence. Part of their obligations was to correct any errors, ambiguities or omissions arising, deal with questions for clarification on design matters from the works director or project director and clarify working drawings where required.

The subcontractor Cementation Bachy was responsible for the design and construction of a diaphragm wall. The design information unfortunately included errors. The judge concluded that Kenchington Ford should have checked and discovered the errors and hence were in breach of their duty.

Mowlem, the main contractor, levied a claim resulting from the incorrect design. The claim was settled and London Underground sought to recover some of the amounts paid to Mowlem from Kenchington Ford. The claim submitted by Mowlem was on a global basis and failed to provide proper details of losses alleged to have been incurred. Judge Wilcox was not impressed by this lack of detail but he did find that Kenchington Ford were obliged to make a payment to London Underground, although in a substantially smaller amount than was claimed.

9.2.7   Employers who reach agreement with contractors to pay out in respect of claims have major hurdles to jump if they are to recover the sums paid from the architect/engineer. In the first instance it is necessary for them to show that the late issue of design information constituted a breach of duty. Further the employer must be able to demonstrate to the satisfaction of the court that the amount paid in respect of the claim can be linked to that breach of duty.

Settlement of a contractor's claim is often the basis of a legal action for the recovery of the amount paid by the employer from the architect/engineer. To be successful in such an action the employer must to be able to demonstrate that the amount of the settlement was reasonable: *Biggin* v. *Permanite* (1951), *P & O Developments Ltd* v. *Guy's and St Thomas' National Health Service Trust* (1998).

SUMMARY

Employers who pay out claims to contractors for late issue of design information do not automatically have a right to recover those sums from the architect/engineer. It is necessary for the employer to demonstrate a breach of duty by the architect/engineer and that the sums paid result from that breach of duty. In addition the employer will have to show that the amounts paid out were reasonable.

9.3 Will a contractor or subcontractor substantially prejudice its case for additional payment if it fails to keep adequate accurate records?

9.3.1 Max Abrahamson in his book *Engineering Law and The ICE Contract* wrote, 'A party to a dispute, particularly if there is an arbitration will learn three lessons (often too late): the importance of records, the importance of records and the importance of records'. This quotation is relevant to the case of *Attorney General for the Falkland Islands* v. *Gordon Forbes Construction (Falklands) Limited* (2003). A contract was let for the construction of the infrastructure of the East Stanley Housing Development in the Falkland Islands using the FIDIC 4th Edition conditions. These conditions, like most standard forms, provide a procedure which the contractor is required to follow in the event of a claim being submitted. Clause 53.1 states that the contractor, if he intends to claim any additional payment, must give a written notice of his intention to the engineer within 28 days after the event giving rise to the claim. In addition, the contractor is obliged by clause 53.2 to keep such contemporary records necessary to support the claim. The contract goes one stage further under clause 53.3 in requiring the contractor to send to the engineer within 28 days of the written notice a detailed account of the claim. If the contractor fails to send in a written notice and detailed account there is a fall-back position in clause 53.4 which states that in the event of the contractor failing to comply with what after all are very simple obligations, his entitlement to payment will not exceed the amount the engineer or an arbitrator could consider verified by contemporary records.

9.3.2 In the case mentioned, it would seem that the contractor, in considering that he was entitled to additional payment, made his first mistake in failing to send a written notice and detailed account as required by clauses 53.1 and 53.3. The question of whether there were sufficient contemporary records to satisfy the requirements of the fall-back position became the subject of a dispute which was referred to arbitration and subsequently to the court. It would seem that the contemporary records were somewhat lacking and did not prove sufficiently convincing as far as the engineer was concerned to merit payment. The contractor felt this approach unjust as he considered he was able to fill in the gaps where the contemporary records were lacking with verbal testimony from those within his organisation able to provide the necessary information. It was the engineer's view that this type of verbal evidence did not amount to contemporary records. This proved somewhat critical as both parties were agreed that if the contractor failed to comply with these conditions his rights to payment were lost and therefore everything hung on the matter of contemporary records.

9.3.3 The judge considered that the requirements of the contract set out a clear and ordered way of dealing with any claim for an additional payment. Claims have to be notified at the time they arise, contemporary records have to be kept and regular accounts rendered. The whole contractual system is aimed at the early resolution of any queries at the time the claim arises. He then had to decide what was meant by the word contemporary and felt there was little difference between contemporary and the word contemporaneous, i.e.

occurring at the same time. Examples were given by the judge to illustrate the point. If there was a dispute as to whether labour was employed on site during a particular date it would be acceptable to produce time sheets showing the hours worked or even a statement by the contractor indicating that this is what had taken place. It would be exceptional if the records were made more than a few weeks after the event took place. The judge posed something of a conundrum. What would the position be if the contractor wished to claim for labour costs for a four-week period? He had time sheets for weeks 1, 2 and 4 but the records for week 3 were missing. Could the court accept verbal evidence concerning week 3; the judge felt that such verbal evidence would not be acceptable.

9.3.4   The judge found in favour of the authority in deciding that to constitute contemporary records these must be prepared at or about the time of the events which give rise to the claim. Where there is no contemporary record the claim will fail. Where there are contemporary records to support part of the claim, it may succeed on that part of the claim which is supported by the contemporaneous records but not otherwise. If, however, the contemporary records are ambiguous or unclear as to their meaning, then verbal evidence will be accepted to provide clarification. The contractor, probably unaware of Max Abrahamson's words of wisdom, found out when it was too late the importance of keeping proper records.

SUMMARY

Where a contract provides a procedure to be followed, if the contractor or for that matter the subcontractor intends to claim additional payment, a failure to comply will result in a loss of the entitlement. Where records are to be contemporaneous or submitted within a specified time scale, it is crucial for them to be properly made at the time required.

9.4   When a contractor/subcontractor fails to serve a proper claims notice or does not submit details of the claim as required by the contract, can the architect/engineer legitimately reject the claim?

9.4.1   Disputes often arise between contractors or subcontractors and the employer's consultants concerning the service of a written notice in relation to a right to additional payment. Does a lack of a written notice lose the contractor or subcontractor his rights. In other words is the procedure which has been written into the contract a condition precedent to the rights which are provided by the terms of the contract.

9.4.2   Some standard forms of contract make it clear how the lack of appropriate written notice will affect the contractor's entitlements.

The ICE 6th Edition clause 52(4)(e) and the 7th Edition clause 53(5) deal with the problem of late notice and state:

'If the Contractor fails to comply with any of the provisions of this Clause [notice of claims/additional payments] in respect of any claim which he shall seek to make then the Contractor shall be entitled to payment in respect thereof only to the extent that the Engineer has not been prevented from or substantially prejudiced by such failure in investigating the said claim.'

GC/Works/1 1998 condition 46 states that prolongation or disruption costs will not be paid unless the contractor, immediately upon becoming aware that the regular progress of the works has been or is likely to be disrupted or prolonged, gives a notice to the project manager specifying the circumstances.

9.4.3 *Hudson's Building and Engineering Contracts* 11th edition at paragraph 4.132 deals with the matter in general terms when it states:

'Since the purpose of such provisions is to enable the owner to consider the position and its financial consequences (by cancelling an instruction or authorising a variation, for example, he may be in a position to reduce his financial liability if the claim is justified), and since special attention to contemporary records may be essential either to refute or regulate the amount of the claim with precision, there is no doubt that in many if not most cases the courts will be ready to interpret these notice requirements as conditions precedent to a claim, so that failure to give notice within the required period may deprive the contractor of all remedy.'

9.4.4 In the case of *London Borough of Merton* v. *Stanley Hugh Leach* (1985) Mr Justice Vinelott had this to say with regard to the need for a loss and expense notice under clauses 11(6) and 24(1) of JCT 63:

The common features of subclauses 24(1) and 11(6) are first that both are 'if' provisions (if upon written application being made, etc.) that is, provisions which only operate in the event that the contractor invokes them by a written application . . .

He then went on to consider the amount of detail which must be included with the notice:

The question of principle is whether an application under clauses 24(1) or 11(6) [of JCT 63] must contain sufficient information to enable the architect to form an opinion on the questions whether (in the case of clause 24) the regular progress of the work has been materially affected by an event within the numbered sub-paragraphs of clause 24 or (in the case of clause 11(6)) whether the variation has caused direct loss and/or expense of the kind there described and in either case whether the loss and/or expense is such that it would not be reimbursed by payment under other provisions of the contract or in the case of 11(6) under clause 11(4).

The judge pointed out that it would not necessarily be enough simply to make what might be described as a 'bare' application which would satisfy the requirements of clause 11(6) or clause 24(1). The application had to be framed with sufficient particularity to enable the architect to do what he was required to do. It follows that the application must therefore contain sufficient detail for the architect to be able to form an opinion as to whether or not there is any loss or expense to be ascertained.

Mr Justice Vinelott also commented upon the circumstances following a contractor's application that satisfied the minimum requirements of clause 11(6) and/or clause 24(1). The architect had been able to form an opinion favourable to the contractor and was then under a duty to ascertain or instruct the quantity surveyor to ascertain, the alleged loss and/or expense. He said:

> The contractor must clearly co-operate with the architect or the quantity surveyor giving such particulars of the loss or expenses claimed as the architect or quantity surveyor may require to enable him to ascertain the extent of that loss or expense; clearly the contractor cannot complain that the architect has failed to ascertain or to instruct the quantity surveyor to ascertain the amount of direct loss or expense attributable to one of the specified heads if he has failed adequately to answer a request for information which the architect requires if he or the quantity surveyor is to carry out that task.

Later he said:

> If [the contractor] makes a claim but fails to do so with sufficient particularity to enable the architect to perform his duty or if he fails to answer a reasonable request for further information he may lose any right to recover loss or expense under [clause 11(6) or clause 24(1)] and may not be in a position to complain that the architect was in breach of his duty.

9.4.5 In the insurance case of *Kier Construction Ltd* v. *Royal Insurance (UK) Ltd* (1992) the insurance policy required the claimant to notify the insurer as soon as possible if there was an occurrence such that in consequence a claim is to be made. The claim should have been made on 12 June 1989 but it was not submitted until 4 July 1989. The claimant lost his rights as the notice was not served 'as soon as possible'.

9.4.6 In *Rees & Kirby Ltd* v. *Swansea City Council* (1984) the court had this to say with regard to the contractor's claim for finance charges as part of the loss and expense claim and the need for reference to be made in the notice to such charges:

> I agree with the judge's construction of clause 11(6) and 24(1) and with his conclusion that the architect can only ascertain and certify the amount of interest charges lost or expended at the date of the application. It is these charges which are the subject of the application; it is these charges which he has power to investigate, ascertain and certify. I respectfully agree with the judge that the architect would be exceeding his powers were he to take into account further financial charges or other losses accruing during these two periods, however long, and such further charges and losses would be recoverable only, if at all, under a subsequent application or subsequent applications – although he might obtain the respondents' approval to waiving the required applications or extending the time for making them.

9.4.7 The case of *Hersent Offshore SA and Amsterdamse Ballast Beton-Waterbouw BV* v. *Burmah Oil Tankers Ltd* (1978) also deals with the question of notice where it was held that a notice given after completion of the work could not be regarded as a notice given before work commenced or as soon thereafter as is practicable.

9.4.8 From the above it can be seen that a contractor or subcontractor who fails to serve a proper claims notice will in all probability lose his rights. Contractors/subcontractors from time to time fail to comply with the contract requirements in relation to notice or some other procedural matter, and are thus prevented from levying a claim under the contract. If the event giving rise to the claim would also give an entitlement to a common law damages claim, e.g. late issue of architect's drawings, the contractor/subcontractor may be able to claim damages for breach of contract.

9.4.9 Clause 26.6 of JCT 98 states that the provisions of this clause are without prejudice to any other rights and remedies which the contractor may possess. Clause 24(2) of JCT 63 was similarly worded and from the judgment of Mr Justice Vinelott in the case of *Stanley Hugh Leach* v. *London Borough of Merton* (1985) it seems clear that a claim under clause 24(2) of JCT 63 is an alternative to a claim under clause 24(1) where he observed:

> But the contractor is not bound to make an application under clause 24(1). He may prefer to wait until completion of the work and join the claim for damages for breach of obligation to provide instructions, drawings and the like in good time with other claims for damage for breach of obligations under the contract. Alternatively, he can, as I see it, make a claim under clause 24(1) in order to obtain prompt reimbursement and later claim damages for breach of contract, taking the amount awarded under clause 24(1) into account.

9.4.10 It would seem that even without the express retention of common law rights they would not be lost. A clause which excluded all rights except those set out in the contract would be required if this end were to be achieved.

This was the situation arose in *Strachan and Henshaw* v. *Stein* (1997) where a contract was let using the MF/1 conditions of contract.

Condition 44.4 states:

> 'Accordingly, except as provided for in the conditions, neither party shall be obliged or liable to the other in respect of any damages or losses suffered by the other which arise out of, under or in connection with the contract or the works'

It was held by the court that a claim for breach of contract outside the provisions of the contract failed due to the wording of condition 44.4.

9.4.11 The failure of a contractor to submit a written notice of a monetary claim entitlement was dealt with in the case of *Maidenhead Electrical Services* v. *Johnson Controls* (1997). A key question was whether this failure affected the contractor's rights to claim damages for breach of contract. The wording of the contract stated:

> 'PAYMENT ... Any claims by the contractor requesting consideration for payments additional to those provided for in the subcontract or in amendments to the subcontract shall be submitted in writing to the company within 10 days of the occurrence from which the claim arises. If no notification of the claim is received by the company within 28 days of such date, then the said claim shall be automatically invalid.'

The defendants argued that this clause applied to payments additional to those provided for in the subcontract and not those for extensions of time and associated monies which were claims under the subcontract. The court's considered opinion was:

> The issue as reformulated focuses on damages for breach of contract and additional monies under the contract. I do not consider that the limitation applies to claims for damages for breach of contract. I do not see that a claim for damages for breach of contract is a claim for payment additional to those provided for in the subcontract. The words are wholly insufficient to exclude liability for damages for breach of contract in the event of failure to comply with those time limits. In my view, the limitation is directed to the case where, for example, the contract rates are insufficient to cover the contractor's costs. The reference to 'amendments' supports this view. Thus, the relevant wording of condition 17 would not operate to exclude a money claim associated with an extension of time, or a disputed amendment.

9.4.12   Not every attempt to make a written notice a condition precedent succeeds. The case of *Amec Process & Energy Ltd* v. *Stork Engineers and Contractors BV* (1999) arose out of the construction of the topside of a floating oil production unit by Stork for Shell. Part of the work was subcontracted by Stork to Amec. In the subcontract conditions a clause appeared which stated that the contract price could only be changed by a variation. Amec could request a variation to the price if any of a prescribed set of circumstances arose. The clause contained several conditions regarding matters such as notice, the keeping of records and the provision of a full evaluation. Failure to comply resulted in Amec losing the right to a price adjustment.

It was held that the clause was not worded with sufficient clarity to remove Amec's right to claim damages for breach. The judge explained that, if a party to a contract is to be deprived of its right to recover damages for breach, the wording should be in clear, unambiguous terms.

SUMMARY

Whether the lack of a proper claims notice and back up details will result in a contractor or subcontractor losing an entitlement to additional payment will depend upon the wording of the contract. Where the contract states that a notice is a condition precedent then a lack of notice will be fatal. The same would apply if the contract is silent on the matter.

Some contracts, for example the ICE 6th and 7th Editions and GC/Works/1 1998, deal expressly with the effect of lack of notice whereas JCT 98 does not.

Loss of a right to claim under the terms of the contract may not affect a right to recover sums for breach of contract unless the contract expressly excludes these rights.

9.5 With a programme shorter than the contract period, can the contractor/subcontractor claim additional payment if, because of the timing of the issue of the architect/engineer's drawings, he is prevented from completing in accordance with the shortened programme?

9.5.1 Contractors and subcontractors, when submitting tenders, are always seeking to gain a competitive edge when pricing their bid. Often they consider that the proposed contract period is generous and they are capable of completing well within it. If the tender price is based upon a period shorter than the proposed contract period a saving on site overheads and head office contribution will normally be achieved. Contractors and subcontractors will usually be reluctant to make the basis of their pricing known at tender stage and it is only when the programme is produced, usually after the contract has been signed, that completion is shown earlier than the contract completion date.

In the event of the architect/engineer failing to issue drawings and details in sufficient time to meet the shortened programme will this give rise to an entitlement on the part of the contractor/subcontractor to be paid any resultant additional costs?

9.5.2 In the case of *Glenlion Construction Ltd* v. *The Guinness Trust* (1987) Glenlion entered into a contract dated 10 July 1981 with The Guinness Trust. The contract incorporated JCT 63. Item 3.13.4 in the bills of quantities provided for the following:

> 'Progress Chart
> Provide within 1 week from the date of possession, a programme chart of the whole of the works, including the works of nominated subcontractors and suppliers and contractors and others employed direct including public utility companies and showing a completion date no later than the Date for Completion. The chart to be a bar chart in an approved form. Forward 2 copies to the Architect, 1 copy to the Quantity Surveyor and keep up to date. Modify or re-draft.'

9.5.3 Disputes were referred to arbitration where the arbitrator made the following decision:

- On a true construction of the contract the contractor was obliged to provide a programme showing completion of the whole of the works no later than the date for completion. Agreement or approval by the architect of the programme should not relieve the contractor of the duty to complete the whole of the works by the date for completion.
- The contractor was entitled to complete the works on a date earlier than the date for completion in the contract.
- When the contractor programmed to complete earlier than the date for completion there was no implied obligation upon the employer his servants or agents to perform the agreement so as to enable the contractor to carry out and complete the works in accordance with the programme. The reason being that the contractor was not obliged to complete early and therefore if there was an implied term it would enforce an obligation on the employer but not the contractor.

9.5.4   It follows that the contractor is entitled to complete the works earlier than the contract completion date and has a right to do so. There is no corresponding duty, however, on the part of the employer to permit him to do so, and in particular to furnish him with information or otherwise positively co-operate so as to enable him to do so. The contractor is merely free from any contractual restraint and may complete earlier. The employer must not prevent him from doing so but this does not mean that the employer is bound to facilitate in a positive way the implementation of the contractor's privilege or liberty.

9.5.5   The *Glenlion* case which found in favour of the employer dealt solely with the issue of drawings necessary to enable the work to be completed. It did not address additional cost and loss and expense, in general where delays occur and in particular where caused by variations, although the judge did venture to say that it was 'unclear how the variation provisions would have applied'.

Reference was made by the judge to similar authorities and, in particular, to *Keating on Building Contracts*. In the edition referred to (the supplement to the 4th edition) it is stated:

> 'Whilst every case must depend upon the particular express terms and circumstances, it is thought that, upon the facts set out in *Wells* v. *Army & Navy Co-operative Society* (1930) the contractor's argument is bad; and that is the case even though the contractor is required to complete "on or before" the contract date ... There is no authority on this point.'

However, in the 5th and 6th editions of *Keating*, the author goes on to say:

> 'Where the programme date is earlier than the Date for Completion stated in the contract, it may be that some direct loss and/or expense may be recoverable on the grounds of disruption. However, provided that the contractor can still complete within the contract period, he cannot recover prolongation costs (*Glenlion Construction* v. *The Guinness Trust*).'

The *Glenlion* decision is likely to be limited in its effect as to the late issue of drawings necessary to carry out and complete the works. Delays to an early completion programme due to variations which create additional cost may not be caught by it.

9.5.6   The South African case of *Ovcon (Pty) Ltd* v. *Administrator of Natal* (1991) also dealt with this matter. In like manner to *Glenlion* the contractor showed completion in eleven months with a contract period of fifteen months. Three months delay to the programme period was caused by the employer. The court refused to award additional prolongation costs saying that if the contractor had taken the contemplated fifteen months these expenses would have been incurred in any event.

9.5.7   In most situations, it is not the *programme* which is relevant. The contractor must show that his *progress* was affected and that he suffered loss and/or expense thereby.

For example JCT 98 clause 26.1 identifies circumstances giving rise to a claim for loss and expense, stating:

> 'If the contractor ... has incurred or is likely to incur direct loss and/or expense ... due to deferment of giving possession of the site ... or because the regular progress

of the Works or of any part thereof has been or is likely to be materially affected ...
the Architect ... shall ascertain ... the amount of such loss and/or expense ...'

9.5.8   Programmes submitted by the contractor under clause 14 (1) of the ICE 6th
and 7th Editions are for acceptance by the engineer. If the contractor submits
a programme which shows early completion and this is accepted by the
engineer the situation could be different. The contractor would, one expects,
apply to the engineer for drawings to be issued to meet the shortened
programme. If the engineer offered no resistance to these requests but issued
the drawings later than requested an arbitrator might consider that the
information had not been issued at 'a time reasonable in all the circumstances' as required by clause 7(4)(a). This would give rise to a contractual
entitlement for additional payment.

9.5.9   The Society of Construction Law Delay and Disruption Protocol 2002 takes a
slightly different approach to the matter. It expresses the opinion that a
contractor will be entitled to levy a claim if unable to complete in accordance
with a shortened programme, provided the employer is aware of the contractor's intention at the time the contract is entered into.

9.5.10   The wording of the Protocol states:

'1.12.1   If as a result of an Employer Delay, the Contractor is prevented from
completing the works by the *Contractor's planned completion date* (being a
date earlier than the contract completion date), the Contractor should in
principle be entitled to be paid the costs directly caused by the Employer
Delay, notwithstanding that there is no delay to the contract completion
date (and therefore no entitlement to an EOT), provided also that at the
time they enter into the contract the Employer is aware of the Contractor's
intention to complete the works prior to the contract completion date, and
that intention is realistic and achievable.

1.12.2   Where the parties have not addressed this issue in their contract, for the
Contractor to have a valid claim, the Employer must be aware at the time
that contract is entered into of the Contractor's intention to complete prior
to the contract completion date. It is not permissible for the Contractor
after the contract has been entered into, to state that it intends to complete
early, and claim additional costs for being prevented from doing so.

1.12.3   The Protocol recognises that the position it takes on this issue might be
thought to conflict with the decision of Judge Fox-Andrews in the (English) Technology and Construction Court in *Glenlion Construction Ltd* v.
The Guinness Trust (1987) 39 BLR 89, where it was held that there was no
implied term of the building contract in question that the Employer in that
case should so perform the contract as to enable the Contractor to complete the works in accordance with a programme that showed the works
being completed before the contract completion date. Providing the
Employer is aware of the Contractor's intention prior to the contract being
entered into, there should be no such conflict. The Protocol considers that,
as a matter of policy, contractors ought not to be discouraged from
planning to achieve early completion, because of the price advantage that
being able to complete early is likely to have for the Employer. But the
potential for conflict reinforces why the issue should be addressed
directly in every contract.'

SUMMARY

The *Glenlion* case and the South African case of *Ovcon* made it clear that the architect has no implied obligation to issue drawings at such time as would enable the contractor to finish work to meet an early completion programme.

There are, however, a number of angles which contractors/subcontractors may wish to consider such as:

- Has the delay been caused by a variation and therefore the pricing should include prolongation costs?
- Would the argument stand a better chance of success if it could be shown that the timing of the issue of drawings affected the progress of the works? JCT 98 provides payment of additional cost resulting from a delay to the progress of the works.
- In the context of a contract let using the ICE 6th or 7th Editions, was the contractor working to a programme accepted by the engineer? If so the engineer might be obliged to certify additional cost on the basis that drawings had not been issued at a time 'reasonable in all the circumstances'.

These matters take a different line from *Glenlion* which dealt merely with the question of whether there was an implied term in a contract requiring the architect to issue drawings to meet an early completion programme.

The Society of Construction Law Delay and Disruption Protocol states that a contractor will have the right to levy a claim if it is unable to complete in accordance with a shortened programme provided the employer is aware of the contractor's intention at the time the contract is entered into.

9.6 Where a contractor submits a programme (which is accepted or approved) showing completion on the completion date written into the contract, must drawings be issued in good time to enable the contractor to carry out the work at the time and in the sequence indicated on the programme?

9.6.1   Most standard forms of contract provide for the contractor to produce a programme. The contractual requirements as to the detail to be provided in the programme vary from contract to contract. Timings for the issue of drawings and details by the architect or engineer are also provided in the standard forms of contract. No link however is expressed in these standard forms between the clauses which provide for the issue of drawings and details and the ones dealing with the contractor's programme.

9.6.2   JCT 98 is basic in its requirements with regard to the issue of a master programme. Clause 5.3.1.2 requires the contractor to provide the programme to the architect as soon as possible after the execution of the contract.

9.6.3   Clause 5.4.1, which is optional relating to the issue of drawings, requires the architect to issue drawings and details in accordance with the information release schedule produced prior to sending out tender enquiries.

Clause 5.4.2 applies if there is no information release schedule and requires the architect to issue drawings and information 'as and when from time to time may be necessary'. The clause goes on to say:

> 'Such provision shall be made … at a time when, having regard to the progress of the Works, or where in the opinion of the Architect Practical completion of the Works is likely to be achieved before the Completion Date, having regard to such Completion Date, it was reasonably necessary for the Contractor to receive such further drawings'.

This type of wording is extremely difficult to digest and must confuse most contractors and employer's consultants. It would seem, however, that the subclause requires the architect to have regard to the date for completion written into the contract when issuing drawings and details and not a date when completion is anticipated to be early.

The clause also deals with the contractor's obligation to request the architect to issue drawings since it continues:

> 'Where the Contractor is aware and has reasonable grounds for believing that the Architect is not so aware of the time when it is necessary for the Contractor to receive such further drawings or details or instructions the Contractor shall, if and to the extent that it is reasonably practicable to do so, advise the Architect of the time sufficiently in advance of when the Contractor needs such further drawings or details or instructions to enable the Architect to fulfil his obligations under [this] clause.'

This is another awkwardly worded sentence. It seems to be saying that if the contractor realises that the architect is not aware when drawings are required the contractor must notify the architect reasonably in advance.

9.6.4   The ICE 6th and 7th Editions under clause 14 require the contractor to issue a programme showing the order in which he intends to carry out the work within 21 days after the award of the contract.

9.6.5   Clause 7(3) in the ICE 6th and 7th Editions is simple in comparison with the JCT requirements for the issue of information. The contractor is to give 'adequate' notice in writing to the engineer of any further drawings or specification required. By clause 7(4), in the event of the engineer failing to issue the drawings and specification at a time which is 'reasonable in all the circumstances', then any additional resultant cost will be reimbursable to the contractor under clause 60.

In the case of *Neodox Ltd* v. *Swinton & Pendlebury Borough Council* (1958) Mr Justice Diplock, in deciding what was meant by 'a time reasonable in all the circumstances' said:

> What is a reasonable time does not depend solely on the convenience and financial interests of the contractor; the engineer is to have a time to provide the information which is reasonable having regard to the point of view of himself and his staff and the corporation, as well as the point of view of the contractor.

SUMMARY

It seems clear that where the standard conditions apply the contractor is not entitled to have information issued solely to suit his programme. The contracts seem to indicate as important:

- any information release schedule required by the contract
- the contractor sending a request for information
- the progress of the works
- a reasonable time for the engineer to produce the information.

No reference is made in the standard forms to the issue of information to suit the programme, whether it has been accepted or otherwise.

However if work is progressing in accordance with the programme and proper written requests for the issue of drawings have been sent to the architect/engineer, then there may be an obligation to issue drawings in response to those requests.

9.7 Is a contractor/subcontractor entitled to recover the cost of preparing a claim?

9.7.1 When a claims situation arises, contractors and subcontractors are invariably put to cost in preparing a submission to go to the architect or engineer. The question often asked is whether the cost is recoverable as part of the claim ascertainment and payment.

9.7.2 It would seem that if the contractor in preparing and submitting the claim is merely carrying out an expressed contractual obligation to submit a claim then there will be no entitlement to reimbursement. For example, clause 52(4)(c) of the ICE 6th Edition and clause 53(4) of the 7th Edition require the contractor to provide 'full and detailed particulars of the amount claimed'.

9.7.3 A further situation may arise where the contractor has complied with these contractual requirements but nonetheless the architect or engineer has failed to ascertain and certify the sums due for payment. The contractor would then be entitled to claim damages for breach of contract.

Vincent Powell-Smith in an article appearing in *Contract Journal* 30 July 1992 had this to say on the matter with regard to claims under JCT 80 where the architect fails to ascertain loss and expense as required by the contract:

> 'If the contractor invokes clause 26 and does what is required, the Architect is under a duty to ascertain or instruct the Quantity Surveyor to ascertain whether loss or expense is being incurred and its amount. This follows from the wording of clause 26.1 which uses the word 'shall' and which thus imposes a duty on the Architect, provided that the Architect has formed a prior opinion that the contractor has been or is likely to be involved in direct loss and/or expense as a result of the specified event(s) and which is not recoverable under any other provisions of the contract.'

9.7.4 There is no doubt that the employer is liable in damages for breach by the architect of this duty and this is so whether the architect is an employee, e.g.

where the employer is a public authority, or, as is more usual, an independent contractor engaged by the employer. Where clause 26.1 says 'the Architect shall' this in effect means 'the Employer shall procure that the Architect shall'. This point is implicit in the reasoning in *London Borough of Merton* v. *Stanley Hugh Leach* (1985).

9.7.5 A contractor when claiming damages for a breach by the architect or engineer in not ascertaining loss and expense will be governed by the rules in *Hadley* v. *Baxendale* (1854). The damages recoverable under these rules are:

- those arising naturally, i.e. according to the usual course of things, from such breach
- such as may reasonably be supposed to have been in the contemplation of both parties at the time they made the contract.

It may be argued that both parties would contemplate that if the architect fails to ascertain loss and expense, and hence is in breach, the parties should have contemplated that the contractor would be put to expense in preparing a fully documented claim which should therefore be recoverable.

The same type of argument would apply under ICE 6th and 7th Editions and GC/Works/1 1998 conditions of contract where the engineer or supervising officer fails to certify sums due arising out of a claim.

9.7.6 There is a precedent for the payment of managerial costs resulting from a breach in the case of *Tate & Lyle Food Distribution* v. *GLC* (1982). In that case Mr Justice Forbes said:

> I have no doubt that the expenditure of managerial time in remedying an actionable wrong done to a trading concern can properly form the subject matter of a head of special damage.

This argument may be extended to cover the cost of claims preparation following a breach.

9.7.7 Managerial time and costs were an issue in the case of *Euro Pools* v. *Clydesdale Steel Fabrications Ltd* (2003). Euro Pools entered into a contract with the main contractor in connection with work at the swimming baths at Burntisland and East Moseley in Surrey. Filtration equipment supplied by Clydesdale to Euro Pools proved to be faulty. Part of Euro Pool's claim for breach of contract included cost in respect of time input by the managing director and an engineer which was contested by Clydesdale.

Clydesdale's case was that Euro Pool's entitlement was to recover any actual cost or loss which resulted as a direct consequence of the defective equipment. The managing director and engineers were not employed specifically in relation to the remedial works and therefore their salaries would have been paid in any event. This being the case, Euro Pool had sustained no ascertainable loss as a result of the defects. It was accepted, however, that fuel costs in relation to visits to the swimming pools were an extra cost which could be recovered. The judge was not impressed, his view being that a managing director will normally be expected to devote the whole of his time in the affairs of his employer involving company strategy, searching for new markets and the developments of new products. If he is diverted from these tasks to deal with remedial equipment obtained from a supplier, this will deny him the time which he would otherwise have spent on his normal tasks and as such represents a loss to his employer. The same principle applies to engineers in that time spent on remedial work represents time which could have been profitably spent on their employer's work.

The claim included the basic salaries of the managing director and engineers, plus a mark-up for overheads. The mark-up for the engineers was 350% added to the hourly charge with a 50% increase on top of the managing director's salary costs. The percentage addition for the engineers was said to be the usual mark-up for daywork in the industry and the addition in respect of the managing director was intended to represent a reasonable amount. The judge accepted in principle that overheads are rightfully included in the claim. He felt that a contribution should be made to cover such matters as the cost of maintaining business premises, computer systems, head office and accounting staff, salesmen, estimators and general technical support.

9.7.8 The decision in *Lomond Assured Properties* v. *McGrigor Donald* (2000) and *Pegler* v. *Wang* (2000) involved claims in respect of managers who spent time investigating breaches of contract. In neither case were there any records to support time included in the claim. Oral evidence, however, was accepted by the court to support the time which formed the basis of the claim.

9.7.9 Where matters are referred to arbitration an arbitrator has a discretion to direct by whom and to whom costs shall be paid. The exercise of the arbitrator's discretion is limited to costs connected with or leading up to the arbitration. Normally the arbitrator will award costs which have been incurred after the service of the arbitration notice in favour of the successful party. However if costs incurred before the service of the notice are in contemplation of the arbitration then the arbitrator may include them in his award of costs. It may be argued that costs of preparing a claim document which ultimately forms part of the pleadings but is prepared before the arbitration notice is served falls into the category of costs in contemplation of arbitration. A note on the file before the claim is prepared to the effect that it is being prepared in contemplation of arbitration may prove helpful.

9.7.10 There is a reported case where a court had to decide whether a claims consultant's fees should be reimbursed to a successful claimant, *James Longley & Co Ltd* v. *South West Regional Health Authority* (1983). The case arose out of a dispute concerning a successful claimant's right to recover the costs of employing a claims consultant as part of the costs of the action.

An arbitration between the parties was settled after the hearing had lasted sixteen days.

The claimant's bill of costs contained an item of £16 022 for the fees of a claims consultant. It was directed that the fees in so far as they related to work done in preparation of the claimant's final account and to work as a general adviser to the claimants were to be disallowed but allowance was made for £6452 in respect of work done in preparing the claimant's case for arbitration, namely the preparation of three schedules annexed to the Points of Claim.

This case is no authority for the proposition that costs incurred in preparing claims are always recoverable, but rather the contrary since only the costs directly applicable to the arbitration were allowed – less than half the total.

9.7.11 The Society of Construction Law Delay and Disruption Protocol 2002 offers the contractor hope of recovering the cost of preparing a claim in the following terms:

> '1.20.1 Most construction contracts provide that the Contractor may only recover the cost, loss and/or expense it has actually incurred and that this be demonstrated or proved by documentary evidence. The Contractor should not be entitled to additional costs for the preparation of the information, unless it can show that it has been put to additional cost as a result of the unreasonable actions or inactions of the CA in dealing with the Contractor's claim. Similarly, unreasonable actions or inactions by the Contractor in prosecuting its claim should entitle the Employer to recover its costs. The Protocol may be used as a guide as to what is reasonable or unreasonable.'

SUMMARY

It would seem that when approaching the matter of recovery of the costs of preparing a claim, a number of questions should be addressed:

(1) It appears unlikely that, in the absence of express terms in the contract which give an entitlement to payment, the cost of producing documents in support of a claim as required by the conditions of contract will be recovered. In providing this information the contractor or subcontractor is merely complying with the requirements of the contract.

(2) Where the conditions of contract require the architect or engineer, having received notice and details from the contractor or subcontractor, to ascertain loss and expense, any failure to so ascertain will constitute a breach of contract. The cost of further preparation of a claim if it results from the breach may well be recoverable.

(3) If it can be shown that, prior to the service of an arbitration notice, the preparation of the claim is in contemplation of such arbitration, the arbitrator *may*, in exercising a discretion with regard to the award of costs, include the cost of preparing the claim. The courts have now accepted as part of a claim for breach of contract the cost of managerial time spent in investigating the breaches.

9.8 Will the courts enforce claims for head office overheads based upon the *Hudson* or *Emden* formulae or must the contractor be able to show an increase in expenditure on head office overheads resulting from the overrun?

9.8.1 Contractors and subcontractors when submitting claims for prolongation will normally include an item for head office costs. This can usually take one of two forms.

Firstly the argument may be that the contractor, in having resources locked into a site during the prolongation period, has lost the opportunity of using those resources on other sites where they would have earned a contribution to the costs of running the head office. For this argument to be successful the contractor or subcontractor would have to show that work was reasonably plentiful and that on a balance of probabilities other work was or would have been available. This is sometimes referred to as 'unabsorbed head office overheads' in that the level of head office cost continues but the revenue stream from the particular contract suffering the delay shows a shortfall.

If the contractor is unable to demonstrate a loss of opportunity he may be able to demonstrate that time and cost of identified resources at head office have been incurred during the overrun period.

Any head office overheads recovered through extra works and the expenditure of provisional sums should be credited against sums otherwise claimed.

9.8.2 The loss of opportunity method of claiming for unabsorbed overheads has been converted into three alternative formulae, Hudson, Emden and Eichleay.

9.8.3 *Hudson's Building and Engineering Contracts*, 11th edition at paragraph 8.182 sets out the famous 'Hudson Formula' with the head office percentage used in the calculation of the contract price:

$$\frac{\text{Head Office Percentage}}{100} \times \frac{\text{Contract Sum}}{\text{Contract Period (in weeks)}} \times \text{Period of Delay (weeks)}$$

An alternative formula is produced in *Emden's Building Contracts and Practice*, 6th edition Volume 2, page N/46:

$$\frac{h}{100} \times \frac{c}{cp} \times pd$$

h = head office percentage – arrived at by dividing the total overhead cost and profit of the contractor's organisation as a whole by the total turnover all extracted from the contractor's year end accounts.

c = contract sum

cp = contract period in weeks

pd = period of delay in weeks

In essence the difference between the two formulae is that Hudson uses the head office percentage used in the calculation of the contract sum whereas in Emden the figures from which the head office percentage is calculated are extracted from the contractor's year end accounts.

9.8.4 In *Ellis-Don Ltd* v. *The Parking Authority of Toronto* (1978) the court held that the plaintiffs had been delayed by $17\frac{1}{2}$ weeks as a result of the defendants' failure to obtain a necessary permit. As a result the plaintiffs were entitled to a weekly sum in respect of costs of overheads and loss of profits which employed as a calculating factor 3.8% of the tender sum, i.e. Hudson.

In *Whittall Builders Co Ltd* v. *Chester-le-Street District Council* (1985), Mr Recorder Percival QC, in illustrating the Emden formula, said:

> Lastly, I come to overheads and profit. What has to be calculated here is the contribution to off-site overheads and profit which the contractor might reasonably have expected to earn with these resources if not deprived of them. The percentage to be taken for overheads and profits for this purpose is not therefore the percentage allowed by the contractor in compiling the price for this particular contract, which may have been larger or smaller than his usual percentage and may or may not have been realised.
>
> It is not that percentage that one has to take for this purpose but the average percentage earned by the contractor on his turnover as shown by the contractor's accounts.
>
> On that basis it is clear to me that the calculation which I have to make is as follows. I start with the figure of 14.15 agreed by the parties as the figure for overheads and profit as a percentage of turnover, and I divide that by 100, then multiply by the contract figure of £404 759 divided by 78 to give the turnover per week, and then multiply by 30; and that gives the figure of £22 028.

9.8.5 In *J.F. Finnegan Ltd* v. *Sheffield City Council* (1988), the contractor claimed *inter alia* payment in respect of additional overheads on a housing improvement contract which was subject to delay. The contract was JCT 63.

Sir William Stabb QC, in finding for the contractor, said:

> It is generally accepted that, on principle, a contractor who is delayed in completing a contract due to the default of his employer, may properly have a claim for head office or off-site overheads during the period of delay, on the basis that the work-force, but for the delay, might have had the opportunity of being employed on another contract which would have had the effect of funding the overheads during the overrun period. This principle was approved in the Canadian case of *Shore & Horwitz Construction Co Ltd* v. *Franki of Canada* (1967), and was also applied by Mr Recorder Percival QC, in the unreported case of *Whittall Builders Company Limited* v. *Chester le Street District Council*. Furthermore, in *Hudson's Building Contracts*, at page 599 of the 10th edition, a simple formula is set out to determine the amount of the loss of funding of overheads and profit during the period of overrun.

The contractor unfortunately had not calculated his head office overhead claim using the Hudson formula but one of his own invention. This led Sir William to comment:

> However, I confess that I consider the plaintiffs' method of calculation of the overheads on the basis of a notional contract valued by uplifting the value of the direct cost by the constant of 3.51 as being too speculative and I infinitely prefer the Hudson formula which, in my judgment, is the right one to apply in this case, that is to say, overhead and profit percentage based upon a fair annual average, multiplied by the contract sum and the period of delay in weeks, divided by the contract period.

However Sir William obviously did not fully understand Hudson's formula which includes the head office overhead and profit percentage upon which the tender was based and not the overhead and profit percentage based upon a fair annual average which is more in line with Emden's formula. Nevertheless judicial approval of the formula approach in appropriate circumstances is clear.

9.8.6 In North America head office overheads are often calculated employing the Eichleay formula. This formula is calculated by comparing the value of work carried out in the contract period for the project with the value of work carried out by the company as a whole for the contract period. A share of head office overheads for the company can then be allocated in the same ratio and expressed as a lump sum to the particular contract. The amount of head office overhead allocated to the particular contract is then expressed as a weekly amount by dividing it by the contract period. The period of delay can then be multiplied by the weekly amount to give a total sum claimed.

Expressed as a formula converted into English from American jargon the formula reads:

$$\frac{\text{Value of contract work during contract period}}{\text{Total value of work for the company as a whole during contract period}} \times \begin{array}{l}\text{Total head office}\\\text{overheads costs}\\\text{expended during}\\\text{contract period}\end{array} = \begin{array}{l}\text{Head office overhead}\\\text{costs allocated to the}\\\text{contract}\end{array}$$

$$\frac{\text{Head office overhead costs allocated to the contract}}{\text{Contract period}} \times \text{Period of Delay} = \text{Amount Claimed}$$

9.8.7 The case of *Property and Land Contractors Ltd* v. *Alfred McAlpine Homes North Ltd* (1997) involved an Eichleay type of calculation in which the court had to decide a contractor's entitlement in respect of the recovery of head office overheads. The dispute arose out of a contract between Alfred McAlpine Homes North Ltd and Property and Land Contractors Ltd for the construction of an estate of 22 homes for sale. JCT 80 conditions as amended applied. Due to poor progress with the sale of the houses an instruction was given to Property and Land Contractors to suspend the works which led to a claim being submitted under clause 26 of the conditions. The claim was not resolved and the matter was referred to arbitration.

One of the items in dispute related to head office overheads. A claim was submitted in the alternative. The first alternative was based upon the application of the Emden formula. It was the contractor's method to only undertake one major project at any one time. A second project at Tollerton was planned and it was agreed that Property and Land Contractors intended to carry out this development for its parent company after completing Shipton. It was claimed that, due to the postponement, completion of the work was delayed from 20 May 1990 until 25 November 1990, and prevented the plaintiff from carrying out the development at Tollerton. The plaintiff claimed that due to the overrun at Shipton he lost an opportunity of carrying out the Tollerton work which would have provided a recovery of overheads.

The Emden formula was employed as a means of calculating the head office overheads. This argument was rejected by the arbitrator on the facts of the case as he was not convinced that the suspension resulted in the plaintiff being unable to work at Tollerton or anywhere else.

In the alternative the plaintiff claimed for the recovery of head office overheads actually expended. The arbitrator was persuaded that the facilities provided by head office were essential to the control and organisation of the project. He considered it irrelevant to make an attempt to show that particular head office costs were increased as a result of this delay. It was evident to him that the head office costs were related to the works for the delay period. The plaintiff's method of calculation was:

> to extract from the company's account the overhead costs excluding fixed costs not related specifically to progress on the site (i.e. directors' remuneration, telephone, staff salaries, general administration, private pension plan, rent, rates, light, heat and cleaning and insurance) to express such annual costs as weekly averages for both 1990 and 1991, and multiply the resulting weekly averages by the period of overrun in each year and thus to produce a figure referred to as 'C'.

The calculation gave the total overheads for the period of delay and had to be allocated between Shipton and other work being undertaken at the same time. This was achieved by a simple formula.

$$\frac{\text{Value of work at Shipton}}{\text{Total value of work at all sites}} \times \text{total overheads (C)} = \text{amount claimed}$$

The formula in essence is the Eichleay formula.

McAlpine's argument was that the arbitrator had erred in law because he had awarded the plaintiff costs which would have been incurred in any event and could not therefore be classed as direct loss and expense.

The court found in favour of the plaintiff and against McAlpine, but had the following reservations.

> All these observations like those of Lord Lloyd in *Ruxley*, of Mr Justice Forbes in *Tate & Lyle*, and of Sir Anthony May in *Keating* all suppose, either expressly or implicitly, that there may be some loss as a result of the event complained of, so that in the case of delay to the completion of a construction contract there will be some 'under recovery' towards the cost of fixed overheads as a result of the reduced volume of work occasioned by the delay, but this state of affairs must of course be established as a matter of fact. If the contractor's overall business is not diminishing during the period of delay, so that where, for example, as a result of an increase in the volume of work on the contract in question arising from variation etc., or for other reasons, there will be a commensurate contribution towards the overheads which offsets any supposed loss, or if, as a result of other work, there is no reduction in overall turnover so that the cost of the fixed overheads continues to be met from other sources, there will be no loss attributable to the delay.

9.8.8   In *St Modwen Developments Ltd* v. *Bowmer & Kirkland* (1996) the arbitrator found in favour of the contractor in awarding head office overheads based upon a formula method of recovery. An appeal was lodged, not because of the use of the formula but on the basis that no evidence had been presented to prove that the contractor was unable to use the head office resources during the period of prolongation to generate profits on other contracts.

The court seemed impressed by the learned work of Mr Duncan Wallace in his 10th edition of *Hudson* where he states:

> 'However, it is vital to appreciate that both these formulae (Hudson and Eichleay) were evolved during the 1960s at a time of high economic activity in construction. Both assume the existence of a favourable market where an adequate profit and fixed overhead percentage will be available to be earned during the delay period. Both also very importantly, assume an element of constraint – that is to say that the contractor's resources (principally of working capital and key personnel, it is suggested) will be limited or stretched, so that he will be unable to take on work elsewhere.'

The court supported the arbitrator on the grounds that both expert and evidence of fact had been heard on which the arbitrator was entitled to base his award. In any event the arbitrator in his terms of appointment had wisely included a clause which allowed him to use his professional knowledge and experience as assistance in determining the matters in dispute. The court felt that the arbitrator would draw on his own experience to conclude from the evidence before him in the manner that he did.

9.8.9 *Amec* v. *Cadmus* (1996) is another example of a claim for the recovery of head office costs. The judge, Mr Recorder Kallipetis QC held:

> It is for the plaintiff to demonstrate that he has suffered the loss he is seeking to recover ... [and] ... this proof must include the keeping of some form of record that the time was excessive and their attention was diverted in such a way that loss was incurred.

He went on to say the plaintiff must:

> place some evidence before the court that there was other work available which, but for the delay, he would have secured ... Thus he is able to demonstrate that he would have recouped his overheads from those other contracts and, thus, is entitled to an extra payment in respect of any delay period awarded in the instant contract.

9.8.10 In the case of *Norwest Holst Construction Ltd* v. *Co-operative Wholesale Society* (1998) the court had to decide whether the arbitrator was entitled to use a formula to ascertain direct loss and/or expense

- in relation to additional overheads and/or
- in relation to unabsorbed overheads (in the absence of a finding by the arbitrator of a reduction of turnover directly attributable to the delay which caused the loss and expense).

A JCT contract applied.

The arbitrator used the Emden formula which he reduced by four fifths when making the award.

It was the court's decision that an Emden-style formula is sustainable when the following circumstances occur:

(1) The loss in question must be proved to have occurred.
(2) The delay in question must be shown to have caused the contractor to decline to take on other work which was available and which would have contributed to its overhead recovery. Alternatively, it must have caused a reduction in the overhead recovery in the relevant financial year or years which would have been earned but for that delay.
(3) The delay must not have had associated with it a commensurate increase in turnover and recovery towards overheads [e.g. a variation].
(4) The overheads must not have been ones which would have been incurred in any event without the contractor achieving turnover to pay for them.
(5) There must have been no change in the market affecting the possibility of earning profit elsewhere and an alternative market must have been available. Furthermore, there must have been no means for the contractor to deploy its resources elsewhere despite the delay. In other words, there must not have been a constraint in recovery of overheads elsewhere.

In this claim by CWS, it is not a situation where CWS need merely establish how it would have acted but for the delay, CWS must, instead, establish a number of inter-related facts:

(1) That its senior management, who spent time on this contract in the period of delay, would have spent that time on other contracts on which CWS was working at the time.
(2) Had that time been spent in that alternative way, the administrative tasks that would have been undertaken on those other contracts would have caused a variety of people and a variety of contractors and suppliers with whom CWS were working to have performed more speedily, economically and efficiently such that CWS' profit from those contracts would have improved. Only some of these individuals and companies would have been within CWS' control.
(3) As a result of CWS' additional profitability, it would have earned a greater contribution to its overheads from these contracts than it actually did.

9.8.11 In the case of *Beechwood Developments Ltd* v. *Stuart Mitchell* (2001) Beechwood undertook to build houses on a site in Bearsden, which is near Glasgow, for Westpoint, a developer. Delays to the commencement of the work resulted from errors made in carrying out the site survey, which was the responsibility of Westbrook. Beechwood's claim for damages fell under three heads: direct site costs, increased cost of materials and loss of contribution to overheads and profits. With regard to the overhead claim, Beechwood argued that the delay resulted in a reduction in the turnover during the financial year in which the delay occurred. That turnover would have contributed to the cost of running the head office and the company's profit. The average overhead and profit for the two previous years as a percentage of turnover was 14.7%. It was claimed that by applying this percentage to the contract sum, the anticipated total sum for overheads and profit which Beechwood anticipated recovering from the project could be ascertained. The weekly amount expected to be recovered was then calculated by dividing this sum by the contract period in weeks. How much of the overrun period could be attributable to the delay at the beginning of the contract was a matter of argument. Beechwood claimed 22 weeks; the judge on the other hand considered that the appropriate period was only 10 weeks. The judge noted that the percentage was derived from Beechwood's accounts and did not represent the overheads built into the contract price. He did not, however, seem to think that this was too important. This, however, does raise an important point. The Hudson formula uses the percentage for overheads that the contractor used in building up his tender figure. Emden's formula uses the overhead figures extracted from the contractor's accounts. Evidence to the effect that Beechwood's gross profit had fallen from £200 000 in the year before the problem at Bearsden down to £9500 in the problem year was also convincing. In his decision the judge awarded a percentage of 11.65% applied to the contract sum and a delay period of 10 weeks. In all, the total of the award under this head amounted to £31 007.

9.8.12 The Society of Construction Law Delay and Disruption Protocol 2002 gives qualified support to the use of the formulae in the following respect:

> '1.16.3 Unless the terms of the contract render unabsorbed overheads irrecoverable, they are generally recoverable as a foreseeable cost resulting from prolongation. The Contractor must be able to demonstrate that because of the Employer Risk Events it was prevented from taking on other overhead-earning work.
>
> 1.16.6 The three most commonly used formulae for assessing unabsorbed head office overheads are Hudson, Emden and Eichleay.
>
> 1.16.7 The use of the Hudson's formula is not supported. This is because it is dependent on the adequacy or otherwise of the tender in question, and because the calculation is derived from a number which in itself contains an element of head office overheads and profit, so there is double counting.
>
> 1.16.8 In the limited circumstances where a head office overhead formula is to be used, the Protocol prefers the use of the Emden and Eichleay formulae. However, in relation to the Eichleay formula, if a significant proportion (more than, say, 10%) of the final contract valuation is made up of the value of variations, then it will be necessary to make an adjustment to the input into the formula, to take account of the fact that the variations themselves are likely to contain a contribution to head office overheads and profit.
>
> 1.16.9 The CA or, in the event of a dispute, the person deciding the dispute, should not be absolutely bound by the results of a formula calculation. It is possible that the use of a particular formula will produce an anomalous result because of a particular input into it. It is suggested that the result of the use of one formula be cross-checked using another formula.
>
> 1.16.10 The tender allowance for head office overheads may be used, if that is what the parties for convenience wish to do.'

9.8.13 What the courts have not yet been asked to address is the extent to which resources are retained on site during the overrun period. The formulae would seem inappropriate if some of the site management personnel during the overrun period were in fact re-allocated to other projects with the exception of a small essential team.

9.8.14 An alternative method of presenting a claim for head office overheads was used in the case of *Tate & Lyle* v. *GLC* (1983). This is not a building case but is of general application. The plaintiffs claimed 2.5% of the prime cost for managerial time, and the judge accepted that such a head of claim was admissible but he did not accept the method of calculation and the application was rejected. He remarked that it was up to managers to keep time records of their activities.

> I have no doubt that the expenditure of managerial time in remedying actionable wrong can properly form the subject matter of a head of special damage. In a case such as this it would be wholly unrealistic to assume that no such additional managerial time was in fact expended. I would also accept that it must be extremely difficult to quantify. But modern office arrangements provide for the recording of time spent on particular projects. I do not believe that it would have been impossible for the plaintiffs in this case to have kept some record to show the extent to which their trading routine was disturbed by the necessity for continual dredging sessions ... While I am satisfied that this head of damage can properly be claimed, I am not prepared to advance into an area of pure speculation when it comes to quantum. I feel bound to hold that the plaintiffs have failed to prove that any sum is due under this head.

The decision in *Babcock Energy* v. *Lodge Sturtevant* (1994) reinforced the contractor's entitlement to recover head office overheads based upon accurately recorded costs. Judge Lloyd in the *Babcock* case being influenced by the *Tate & Lyle* decision.

9.8.15　The courts have also recognised an entitlement to payment of managerial costs associated with investigating breaches of contract in *Euro Pools* v. *Clydesdale Steel Fabrications Ltd* (2003) (see section 9.7.7) and *Lomond Assured Properties* v. *McGrigor Donald* (2000) and *Pegler* v. *Wang* (2000) (see section 9.7.8).

9.8.16　The decisions in *Tate & Lyle, Babcock Energy, Euro Pools* v. *Clydesdale, Lomond Assured* v. *McGrigor Donald* and *Pegler* v. *Wang*, it is submitted, cast doubt on the formula methods of calculation only in so far as the contractor is unable to demonstrate a loss of opportunity to use the resources on other sites.

This would often be the case during periods when the construction industry is in recession. Contractors when faced with a downturn in work often lay off resources as each job finishes until the size of the organisation is suitable for the economic climate current at the time. It would, under these circumstances, be inappropriate to claim reimbursement of head office overheads on the basis of a lost opportunity.

When applying the formula method other matters have to be taken into consideration:

- Credit must be given for any additional overheads recovered by way of the final account.
- Allowance should be made for the situation where resources during a delay period have been substantially reduced and deployed to other sites.
- Credit should be given against the formula calculation for head office staff claimed for elsewhere, e.g. the contracts manager.

SUMMARY

There is no hard and fast rule that a contractor who has an entitlement to the recovery of additional costs due to an overrun to the completion date is entitled to be paid for additional overheads calculated on a formula basis. It will depend in the first instance upon the wording of the contract. The GC/Works/1 1998 conditions refer in condition 46 only to the recovery of expense which would probably exclude any loss of contribution to overheads due to loss of opportunity to use the resources on other projects.

Whether the contractor on other commonly used standard forms of contract would be entitled to payment based on the formula depends upon the contractor's ability to prove that an overrun to the contract period has resulted in the loss of an opportunity to employ the resources on other work. Further the contractor would have to show that the lost overheads had not been recovered by way of variations.

Where it is not appropriate to base a claim on a loss of opportunity which would bring in a formula method the contractor will be left to base its claim on actual head office costs incurred.

Whichever method of calculation is employed the courts will require supporting evidence that either opportunities have been lost or additional costs incurred.

9.9 Where a delay to completion for late issue of information has been recognised, are loss and expense or additional cost claims in respect of extended preliminaries properly evaluated using the rates and prices in the bills of quantities?

9.9.1 Matters such as site supervision, equipment, health and safety, welfare, storage and the like normally fall within the definition of preliminaries.

The time honoured method adopted by quantity surveyors and engineers when evaluating a contractor's overrun claim is to use the preliminaries as priced in the bill of quantities. Contractors and subcontractors have in the past accepted this method with some reluctance. Nowadays contractors and subcontractors are openly asking whether this method is correct.

9.9.2 The wording in the main claims clause in JCT 98, clause 26(1), is: ... 'that he has incurred or is likely to incur direct loss and/or expense'. Other JCT contracts are similarly worded.

The 5th edition of *Keating on Building Contracts* at page 582 has this to say concerning the meaning of direct loss and/or expense:

> 'This was considered by the Court of Appeal in *F G Minter* v. *W.H.T.S.O.* The court held that direct loss and/or expense is loss and expense which arises naturally and in the ordinary course of things, as comprised in the first limb in *Hadley* v. *Baxendale*. The court approved the definition of "direct damage" in *Saint Line Ltd* v. *Richardsons* as "that which flows naturally from the breach without other intervening cause and independently of special circumstances, whereas indirect damage does not so flow". It follows from the decision in *Minter* that the sole

question which arises in relation to any head of claim put forward by a contractor is whether such claim properly falls within the first limb in *Hadley* v. *Baxendale* so that it may be said to arise naturally and in the ordinary course of things.'

9.9.3 A similar line was taken by Mr Justice Megaw in *Wraight Ltd* v. *PH & T Holdings* (1968) when he said:

In my judgment, there are no grounds for giving to the words 'direct loss and/or damage caused to the contractor by the determination' any other meaning than that which they have, for example, in a case of breach of contract or other question of relationship of a fault to damage in a legal context. Therefore it follows ... that the [contractors] are, as a matter of law entitled to recover that which they would have obtained if this contract had been fulfilled in terms of the picture visualised in advance but which they have not obtained.

9.9.4 The ICE 6th and 7th Editions define cost in clause 1(5):

'The word "cost" when used in the Conditions of Contract means all expenditure properly incurred or to be incurred whether on or off the Site including overhead finance and other charges properly allocable thereto but does not include any allowance for profit.'

GC/Works/1 1998 defines expense in condition 46(6):

'Expense shall mean money expended by the contractor, but shall not include any sum expended, or loss incurred, by him by way of interest or finance charges however described.'

MF/1 defines costs in clause 1.1j as:

'all expenses and costs incurred including overhead and financing charges properly allocable thereto with no allowance for profit.'

Cost is defined in the *Concise Oxford Dictionary* as 'Price paid for thing'.

9.9.5 It would seem from the above that actual cost or loss should be the basis on which claims are based and not the preliminaries as priced in the bills of quantities.

SUMMARY

Having said that extended preliminaries should be based on actual cost it is necessary to state that this means reasonable costs which flow from the late issue of drawings. It will be necessary to identify the periods of time which were affected and to include only those preliminary items where extra costs were incurred.

From the definitions in the various contracts and text book references it seems clear that when evaluating loss and expense, expense or cost claims in respect of extended preliminaries, actual cost or loss should be the basis on which the evaluation should be made and not the prices of the preliminaries in the bills of quantities.

9.10 Once it is established that additional payment is due for prolongation resulting from employer delays, should the evaluation relate to the period when the effect of the delay occurs or by reference to the overrun period at the end of the contract?

9.10.1 It is common practice for decisions concerning the award of an extension of time to be made before considering the matter of prolongation costs. Once an extension of time has been granted, the evaluation of the additional prolongation costs is often related to the period between the contract completion date and the extended completion date. Is this the correct method?

The intention of most construction contracts is for the contractor/subcontractor to be reimbursed the additional cost which results from employer delays. This involves a comparison between the actual costs incurred and what the cost would have been had no delay occurred. Where, for example, time is lost awaiting details for external cladding, which causes a four-week delay to the critical path, evaluating the prolongation costs associated with the four weeks on site, following the contract completion date, would obviously not produce the correct answer. A more accurate evaluation would be achieved by reference to the costs incurred during the four weeks when the information was late in arriving.

9.10.2 The Society of Construction Law Delay and Disruption Protocol 2002, as at clauses 1.11.2 and 1.22.3, with regard to this matter, states:

> 'Arguments commonly arise as to the time when recoverable prolongation compensation is to be assessed: is it to be assessed by reference to the period when the Employer Delay occurred (when the daily or weekly amount of expenditure and therefore compensation may be high) or by reference to the extended period at the end of the contract (when the amount of compensation may be much lower)? The answer to this question is that the period to be evaluated is that in which the effect of the Employer Event Risk was felt.'

SUMMARY

Once it is established that additional payment is due for prolongation resulting from employer delays, the evaluation should relate to the period when the effect of the delay occurs and not by reference to the overrun period at the end of the contract.

9.11 When ascertaining contractor's claims on behalf of employers how should consultants deal with finance charges which form part of the calculation of the claim?

9.11.1 Contractors and subcontractors when submitting claims for loss and expense or additional cost will invariably include sums in respect of finance charges. The argument is that they have been stood out of their money for considerable periods of time which has involved borrowing to make up the shortfall. Interest has to be paid to the bank or, if money is taken off deposit, interest is lost.

9.11.2 There is now no doubt that the contractor is entitled to relief, by way of loss and/or expense, for the cost of financing. In his judgement in *FG Minter Ltd v. Welsh Health Technical Services Organisation* (1980) Lord Justice Stephenson said:

> It is further agreed that in the building and construction industry the 'cash flow' is vital to the contractor and delay in paying him for the work he does naturally results in the ordinary course of things in his being short of working capital, having to borrow capital to pay wages and hire charges and locking up in plant, labour and materials capital which he would have invested elsewhere. The loss of the interest which he has to pay on the capital he is forced to borrow and on the capital which he is not free to invest would be recoverable for the employer's breach of contract within the first rule in *Hadley* v. *Baxendale* (1854) without resorting to the second, and would accordingly be a direct loss if an authorised variation of the works, or the regular progress of the works having been materially affected by an event specified in clause 24(1), has involved the contractor in that loss.

9.11.3 In the case of *Rees & Kirby Limited* v. *Swansea City Council* (1985) the court held, in respect of the sum of £206 629 claimed for interest as part of a claim, that the contractor was entitled both legally and morally to every penny. The Court of Appeal confirmed that financing costs were a recoverable head of loss and expense and stated that such costs should be calculated at compound interest, with periodic rests taken into account. However the amount of interest awarded by the judge was reduced to take account of a period when negotiation took place between the employer and contractor in an attempt to settle the dispute.

9.11.4 When ascertaining the cost of financing, the following should be taken into account:

- The appropriate rate of interest is that actually paid by the contractor provided it is not unreasonable. In the event of the contractor paying well above or well below prevailing market rates it seems from *Tate & Lyle* v. *GLC* (1983) that appropriate rates are those 'at which [contractors] in general borrow money'.
- The cost of finance shall be calculated on the basis that it is charged by the contractor's bank, i.e. using the same rates and compounding accrued interest at the same intervals.
- Where the contractor is self-financed or financed from within its corporate group the appropriate rate of interest is that earned by the contractor (or its group) on monies it has placed on deposit.
- Account should be taken of actual negative cash flows by way of primary expense, i.e. expenses are incurred progressively.

9.11.5 *Amec Process & Energy Ltd* v. *Stork Engineers & Contractors No. 4* (2002) is a case where the court was asked to decide upon the matter of finance charges. Amec was Stork's subcontractor on a contract to convert a ship's hull into a floating production platform. In an earlier judgement, Amec had been awarded damages in respect of additional costs incurred due to variations in the cost and for lost hours. Amec was entitled to be paid at 'net cost plus a percentage mark up to cover all costs, overheads and profit in connection with the personnel involved'.

It was held that these costs should include interest paid by Amec to enable work to be financed and for the personnel to be paid, since these finance costs are 'a percentage mark up to cover all costs, overheads...'. The interest awarded was compounded quarterly.

9.11.6 The principle that finance charges should be paid as part of a loss and expense claim is also recognised in Scotland following the decision in *Ogilvie Ltd* v. *City of Glasgow District Council* (1993).

9.11.7 Some standard contract forms expressly refer to finance charges. Under the ICE 6th and 7th Editions 'cost' is defined as 'including overhead finance and other charges'.

MF/1 also defines 'cost' as 'including overhead and finance charges.'

GC/Works/1 1998 deals with finance charges in condition 47(1) in a different manner where it states:

> 'The Employer shall pay the contractor an amount by way of interest or finance charges (hereafter together called "finance charges") only in the event that money is withheld from him under the contract because, either
>
> a.   the Employer, PM or QS has failed to comply with any time limit specified in the Contract or, where the parties agree at any time to vary any such time limit, that time limit as varied, or
> b.   the QS varies any decision of his which he has notified to the contractor.'

Condition 46 (6) excludes finance charges from the definition of expense.

9.11.8 The Society of Construction Law Delay and Disruption Protocol 2002 offers the following advice concerning finance charges:

> *'Interest as damages/finance charges*
> 1.15.4   It is the position in most areas of business that interest payable on bank borrowings (to replace the money due) or the lost opportunity to earn interest on bank deposits, is quantifiable as damages where the claimant can show:
> 1.15.4.1   that such loss has actually been suffered; and
> 1.15.4.2   that this loss was within the reasonable contemplation of the parties at the time of contracting.
> 1.15.5   It is recognised that, in the construction industry, it will always be in the contemplation of the parties at the time they enter into their contract that if deprived of money the Contractor will pay interest or lose the ability to earn interest. Contractors therefore need only establish that the loss was actually suffered.
>
> *Time when interest starts to run*
> 1.15.6   There are often arguments as to the date on which interest on a Contractor's claim should start to run. Contractors will argue that it should be the date on which they incurred expenditure for which they are entitled to compensation. Employers will say that interest should run only from the date that the Contractor has provided all information needed to satisfy them that the expenditure has been incurred.

1.15.7 The appropriate starting date will not be the same in all circumstances, but generally the starting date for the payment of interest should be the earliest date on which the principal sum could have become payable, which will be the date for payment of the certificate issued immediately after the date the Contractor applied for payment of the loss and/or expense. This will be subject to any notice requirements in the contract. In contracts where there are no certificates, the Protocol recommends that interest should start to run 30 days after the date the Contractor suffered the loss and/or expense.'

SUMMARY

It is clear from case law that contractors and subcontractors are entitled to finance charges as part of their loss and expense or additional cost claims. This is reflected in some contractual definitions. The contractor or sub-contractor will, however, lose his entitlement if he fails to submit a proper notice and details required by the terms of the contract. A further restriction upon the contractor's rights may be imposed by the conditions of contract, for example GC/Works/1 1998.

9.12 Is a contractor/subcontractor entitled to be paid loss of profit as part of a monetary claim?

9.12.1 A contractor is entitled to reimbursement of loss of profit – if he can prove that he was prevented from earning profit elsewhere, in the normal course of his business, as a direct result of one or more of the matters referred to in the condition of the relevant contract: *Peak Construction (Liverpool) Ltd* v. *McKinney Foundation Ltd* (1970).

9.12.2 In *Ellis-Don* v. *The Parking Authority of Toronto* (1978) the judge had this to say with regard to loss and profit:

> If a contractor is entitled to damages for loss of income to cover head office overheads, why should he not also be entitled to damages for loss of income that would result in normal profit?

In *Saint Line* v. *Richardsons Westgate* (1940) a claim was submitted for defects in the main engines of a ship. In addition to an entitlement to the cost of replacement parts the court held there was an entitlement to loss of profit whilst deprived of the use of the vessel.

9.12.3 The amount of such profit, even if proved, must not exceed the level normally to be expected. An exceptionally high profit which might have been earned on another project cannot be reimbursed, unless this fact was known to the employer when the contract was entered into: *Victoria Laundry (Windsor) Ltd* v. *Newman Industries Ltd* (1949).

Account must be taken of any additional profit that may have been earned as a result of additional works being priced in accordance with the conditions of the contract.

9.12.4 Where the contractor's employment has been determined under the contract as a result of default by the employer, the contractor is entitled to be reimbursed the amount of profit that he can prove that he would have made on that particular contract had he been allowed to complete the works, less the amount saved because of the removal of his contractual obligation: *Wraight Ltd* v. *PH & T (Holdings) Ltd* (1980).

9.12.5 However some standard forms of contract specifically exclude profit from the definition of cost or expense.

The ICE 6th and 7th Editions in clause 1(5) state that the word cost 'does not include profit'. MF/1 and the IChemE forms are similarly worded.

SUMMARY

Contractors and subcontractors will normally be entitled to claim loss of profit where they can show that due to the circumstances giving rise to the claim they lost the opportunity of making a profit elsewhere. The exception lies with those standard forms of contract such as the ICE 6th and 7th Editions, MF/1 and the IChemE forms where profit is specifically excluded.

9.13 Is a contractor/subcontractor entitled to be paid acceleration costs as part of his monetary claim?

9.13.1 Contracts frequently fall behind for various reasons leaving the completion date in jeopardy. If the contractor does not voluntarily accelerate the works it will need a separate agreement between the employer and contractor or an express term of the contract to facilitate acceleration.

Clause 46(1) of the ICE 6th and 7th Editions for example gives the engineer power to:

> 'notify the Contractor in writing and the Contractor shall thereupon take such steps as are necessary and to which the Engineer may consent to expedite the progress so as substantially to complete the Works or such Section by that prescribed time or extended time.'

This clause only applies where, due to any reason which does not entitle the contractor to an extension of time, the rate of progress of the works or any section is, in the opinion of the engineer, too slow to comply with the time for completion. This is a very useful provision from the employer's point of view. In the absence of such a clause a contractor in danger of overrunning the completion date may opt to pay liquidated damages in preference to acceleration costs.

9.13.2 A different situation arises where the employer wants to bring forward the contract completion date.

The ICE 6th and 7th Editions include in clause 46(3) for the employer or engineer to request the contractor to complete the works in a time less than the contract period or extended contract period. If the contractor concurs special terms and conditions of payment will have to be agreed.

The JCT Management Form includes detailed provisions for acceleration. In like manner to clause 46(3) of the ICE conditions the provision can only be operated with the agreement of the contractor. Condition 38 of GC/Works/1 1998 operates in a similar fashion.

The Engineering and Construction Contract (NEC) under clause 36 gives the project manager the power to instruct the contractor to submit a quotation for an acceleration to achieve completion before the completion date.

9.13.3 One of the difficulties of operating an acceleration clause is proving the cost of additional resources and reduced outputs which result from acceleration measures. It will be necessary to isolate the costs of acceleration measures. To do so effectively will require the contractor to demonstrate the cost of resources which would have been employed had no acceleration measures been taken.

9.13.4 Contractors and subcontractors often argue that they have been forced to accelerate the works to overcome delays caused by the architect or engineer. There may be some confusion between an acceleration claim and a loss of productivity claim. Normally, in the absence of an instruction to accelerate, the contractor is not entitled to decide unilaterally to accelerate and expect the employer to pay the costs.

The contractor or subcontractor may, however, argue that he chose to accelerate faced with the architect/engineer's refusal or neglect to grant a proper extension of time. This is sometimes referred to as a constructive acceleration order.

Constructive acceleration is defined by the US Corps of Engineers as;

'An act or failure to act by the Employer which does not recognise that the contractor has encountered excusable delay for which he is entitled to a time extension and which required the contractor to accelerate his programme in order to complete the contract requirements by the existing contract completion date. This situation may be brought about by the Employer's denial of a valid request for a contract time extension or by the Employer's untimely granting of a time extension.'

9.13.5 In the Australian case of *Perini Pacific* v. *Commonwealth of Australia* (1969) Mr Justice Macfarlane in the Commercial Court of New South Wales indicated clearly that this type of claim could only be on the basis of some proven breach of contract by the owner – coupled, of course, with proof of damages in the form of completion to time by expenditure greater than would otherwise have been incurred. In that case, the breach consisted of a refusal or failure by the certifier to give any consideration at all to the contractor's applications.

In the case in British Columbia of *Morrison-Knudsen* v. *BC Hydro & Power* the owner, unknown to the contractor, had secretly agreed with a government representative that no extension of time would be granted in view of the pressing need for electricity by the contract completion date. All requests for an extension of time were refused and the contractor carried on and managed to finish the project with only slight delay. The Court of Appeal of British Columbia held that the contractor could have rescinded on the basis of a fundamental breach of contract had he known the real reason for the refusals and, in that event, would have been entitled, on the basis of established case law, to put his claim alternatively on a quantum meruit basis and so to escape from the original contract prices. The fact that the contractor had not rescinded but had completed the project limited his remedy to the usual one of damages, measured in this case in terms of the additional expense incurred in completing to time, i.e. in accelerating progress.

Similarly in *W. Stephenson (Western) Ltd* v. *Metro Canada Ltd* (1987) the Canadian Court held that, by stating that in no circumstances would the contractor receive an extension of time for completion, the employer had deliberately breached the contract and it awarded damages calculated by reference to the contractor's consequential acceleration.

9.13.6   The English courts have been a little slow at recognising a situation where a claim for constructive acceleration would be relevant. However, the situation has been changed by the decision in *Motherwell Bridge Construction Ltd* v. *Micafil* (2002). In this case Motherwell Bridge was a subcontractor to Micafil in connection with the construction of an autoclave for BICC in Kent. Additional welding was required by Micafil which involved a substantial amount of additional man hours. Micafil failed to grant an appropriate extension of time. In an effort to complete by the original completion date, Motherwell Bridge incurred substantial additional costs which formed the subject of a claim.

The matter was referred to court which found in favour of Motherwell Bridge. Judge Toulman considered that Motherwell Bridge was entitled to an extension of time based upon the period of delay which would have been incurred using the original labour force, and as Micafil failed to grant an extension of time, Motherwell Bridge was entitled to be paid the acceleration costs.

9.13.7   It would seem that to recover payment in respect of constructive acceleration a contractor or subcontractor must be able to demonstrate the following:

(1)   An entitlement to an extension of time
(2)   Proper notice as required by the contract has been served
(3)   The employer or his agent has failed to grant an appropriate extension of time
(4)   The employer has requested completion by the original completion date
(5)   The contractor has incurred additional costs due to taking acceleration measures
(6)   The cost to the contractor in taking acceleration measures when estimated in advance of the decision being made to accelerate, is less than the cost resulting from an overrun to the completion date.

SUMMARY

Whether or not a contractor or subcontractor can be required to accelerate the works will depend upon the terms of the contract. The ICE conditions give the engineer power to instruct the contractor to accelerate where, due to his own default, progress is getting behind. Where the employer requires the contractor to accelerate to overcome delays not of his making, agreement between contractor and employer is required.

If the contractor is delayed by the employer , architect or engineer and no recognition is given to the contractor's rights to an extension of time, then a claim for constructive acceleration may be successful if the contractor decides on acceleration as being less expensive than an overrun.

9.14 Where a written claims notice is required to be submitted within a reasonable time, how much time must elapse before the claim can be rejected as being too late?

9.14.1 It is common when drafting contracts where written notice is required, to include an express provision to the effect that the notice must be submitted within a reasonable time. What constitutes a reasonable time is often the subject of hot debate and usually depends upon all the circumstances of the case.

9.14.2 Mr Justice Vinelott in *London Borough of Merton* v. *Stanley Hugh Leach Ltd* (1985) had to decide a number of preliminary issues, one of which related to the written notice requirements of JCT 63 with regard to a contractor's claim for the recovery of direct loss and expense. These state that the application must be made within a reasonable time. Mr Justice Vinelott said that it must not be made so late that the architect can no longer form a competent opinion on the matters which he requires to satisfy himself that the contractor has suffered the loss and expense claimed. However, in considering whether the contractor has acted reasonably it must be borne in mind that the architect is no stranger to the project and it is always open for the architect to call for further information either before or in the course of investigating the claim.

9.14.3 In the insurance case of *Kier Construction Ltd* v. *Royal Insurance (UK) Ltd* (1992) the insurance policy required the claimant to notify the insurer as soon as possible if there was an occurrence such that in consequence a claim is to be made. The claim should have been made on 12 June 1989 but it was not submitted until 4 July 1989. The claimant lost his rights as the notice was not served 'as soon as possible'.

9.14.4 The case of *Hersent Offshore SA and Amsterdamse Ballast Beton-Waterbouw BV* v. *Burmah Oil Tankers Ltd* (1978) also deals with the question of notice where it was held that a notice given after completion of the work could not be regarded as a notice given before work commenced or as soon thereafter as is practicable.

Clause 52 of the contract, amongst other things, provided that:

> 'no increase of the contract price under sub-clause (1) of this clause or variation of rate or price under sub-clause (2) of this clause shall be made unless as soon after the date of the order as is practicable and in the case of extra or additional work before the commencement of the work or as soon thereafter as is practicable notice shall have been given in writing...'

The dispute as to the claimants' entitlement to additional payment in respect of the variation was referred to arbitration. The arbitrator, having found the facts as set out above, determined in his award (amongst other things) that the notice of intention to claim should have been given as soon after the date of the order as was reasonably practicable, that it had not been so given and that accordingly the respondents were not liable to the claimants in respect of the variation.

SUMMARY

There is no hard and fast rule as to when a notice which is required to be submitted within a reasonable time can be rejected as being out of time. What is reasonable depends upon the circumstances.

9.15   What methods of evaluating disruption have been accepted by the courts?

9.15.1   One of the most difficult items to evaluate with any accuracy is disruption. The problem is usually caused by a lack of accurate records.

With the spotlight on linking cause and effect having been created by the decision in *Wharf Properties* v. *Eric Cumine* (1991), claims for disruption will come under greater scrutiny. It is unlikely that contractors and sub-contractors will succeed where their claims for disruption are based simply upon the global overspend on labour for the whole of the contract. More detail will have to be given which isolates the cause of the disruption and evaluates the effect.

The courts have given some assistance in the manner in which disruption should be evaluated.

9.15.2   Comparison of output is one method

In the case of *Whittall Builders Company Ltd* v. *Chester-le-Street District Council* (1985) difficulties were experienced by the employer in giving possession of dwellings on a rehabilitation scheme. The court found that during the period when these problems existed the contractor was grossly hindered in the progress of the work and as a result ordinary and economic planning and arrangement of the work was rendered impossible. However a stage was reached in November 1974 when dwellings were handed over in an orderly fashion and no further disruption occurred. The court had to decide upon the appropriate method of evaluating disruption. Mr Recorder Percival QC in his judgment had this to say:

> Several different approaches were presented and argued. Most of them are highly complicated, but there was one simple one – that was to compare the value to the contractor of the work done per man in the period up to November 1974 with that from November 1974 to the completion of the contract. The figures for this comparison, agreed by the experts for both sides, were £108 per man week while the breaches continued, £161 per man week after they ceased.
>
> . . .
>
> It seemed to me that the most practical way of estimating the loss of productivity, and the one most in accordance with common sense and having the best chance of producing a real answer was to take the total cost of labour and reduce it in the proportions which those actual production figures bear to one another – i.e. by taking one-third of the total as the value lost by the contractor.

> I asked both [Counsel] if they considered that any of the other methods met those same tests as well as that method or whether they could think of any other approach which met them better than that method. In each case the answer was 'no'. Indeed, I think that both agreed with me that that was the most realistic and accurate approach of all those discussed. But whether that be so or not, I hold that that is the best approach open to me, and find that the loss of productivity of labour, and in respect of spot bonuses, which the plaintiff suffered is to be quantified by adding the two together and taking one-third of the total.

There should be little difficulty in calculating the productivity per man week using the build up for interim certificates. The productivity per man week could be calculated by dividing the number of man weeks worked during a valuation period into the value of work carried out, giving an output per man week.

9.15.3 Additional labour and plant schedules provide an alternative route.

It may not be possible due to the nature of the disrupting matters and the complexity of the project to employ the simple approach used in the *Whittall Builders* case. Therefore it may be appropriate to attempt to isolate the additional hours of labour and plant which results from the event giving rise to disruption.

A schedule of matters causing disruption should be prepared. James R. Knowles has devised a schedule JRK/DS/1 which would be appropriate for the task. This appears as Appendix 1 in this book.

9.15.4 Difficulties may be encountered in isolating the additional hours of labour and plant which result from each and every disrupting matter. Inevitably some form of assessment will be necessary.

The courts again have provided assistance in dealing with this problem. In the case of *Chaplin* v. *Hicks*, heard as long ago as 1911, it was held:

> Where it is clear that there has been actual loss resulting from the breach of contract, which it is difficult to estimate in money, it is for the jury to do their best to estimate; it is not necessary that there should be an absolute measure of damages in each case.

Two years later, Justice Meredith in the Canadian case of *Wood* v. *The Grand Valley Railway Co* (1913) had this to say:

> It was clearly impossible under the facts of that case to estimate with anything approaching to mathematical accuracy the damages sustained by the plaintiffs, but it seems to me to be clearly laid down there by the learned judges that such an impossibility cannot 'relieve the wrongdoer of the necessity of paying damages for his breach of contract', and that on the other hand the tribunal to estimate them whether jury or judge, must under such circumstances do 'the best it can' and its conclusion will not be set aside even if 'the amount of the verdict is a matter of guess work'.

The Canadian case of *Penvidic Contracting Co Ltd* v. *International Nickel Co of Canada Ltd* (1975) also provides some guidance on the manner in which disruption should be evaluated where anything like an accurate evaluation is impossible. The dispute arose out of a construction agreement to lay ballast and track for a railroad. The owner was in breach in several respects of its obligation to facilitate the work. The contractor, who had agreed to do the work for a certain sum per ton of ballast, claimed by way of damages the difference between that sum and the larger sum that he would have demanded had he foreseen the adverse conditions caused by the owner's breach of contract. There was evidence that the larger sum would have been a reasonable estimate.

At the hearing, damages were awarded on the basis claimed, but on appeal to the Court of Appeal this portion of the award was disallowed. On further appeal the Supreme Court of Canada held, restoring the trial judgment, that where proof of the actual additional costs caused by the breach of contract was difficult it was proper to award damages on the estimated basis used at trial. The difficulties of accurate assessment cannot relieve the wrongdoer of the duty of paying damages for breach of contract.

9.15.5 The Society of Construction Law Delay and Disruption Protocol 2002 recommends a technique which they refer to as 'the Measured Mile'. This technique was used in the *Whittall Builders* case referred to in 9.15.2. 'The Measured Mile' is described in the Protocol as follows:

> '1.19.7 The starting point for any disruption analysis is to understand what work was carried out, when it was carried out and what resources were used. For this reason, record keeping is just as important for disruption analysis as it is for delay analysis. The most appropriate way to establish disruption is to apply a technique known as 'the Measured Mile'. This compares the productivity achieved on an un-impacted part of the contract with that achieved on the impacted part. Such a comparison factors out issues concerning unrealistic programmes and inefficient working. The comparison can be made on the man hours expended or the units of work performed. However, care must be exercised to compare like with like. For example, it would not be correct to compare work carried out in the learning curve part of an operation with work executed after that period.'

SUMMARY

Disruption can often be difficult to evaluate properly. One of the most satisfactory methods is by comparison of outputs when work is disrupted with outputs when no disruption is taking place, sometimes referred to as 'the Measured Mile'.

In the absence of this type of information records showing the disruption from individual disrupting activities may be illustrated in schedule form.

Finally, when all else fails, courts have accepted a claim for disruption based upon assessed costs.

9.16   Can a claims consultant be liable for incorrect advice?

9.16.1 A claims consultant when offering professional services will be governed by the terms which are expressed or implied in his conditions of appointment. In the absence of any express term to the contrary a claims consultant will have an implied duty to exercise reasonable skill and care: *Bolam* v. *Friern*

Hospital Management Committee (1957) (See 1.1). If there is a failure to exercise such reasonable skill and care and as a result the client suffers loss then the claims consultant will have a liability to the client.

9.16.2 This was illustrated in the case of *Cambridgeshire Construction Ltd* v. *Nottingham Consultants* (1996). The plaintiff contractors retained the defendant claims consultants in relation to a building contract on which they were working. They informed the consultant that they had received the final certificate, dated 13 October 1992, on 19 October 1992. Notice of arbitration had to be given within 28 days and, under JCT 80 clause 30, the certificate became final and conclusive in the absence of an arbitration notice.

It was not until 11 November that the defendants advised the plaintiffs concerning the service of an arbitration notice. The plaintiffs claimed damages for professional negligence, as their right to go to arbitration had been lost. The defendants argued that the 28-day period ran from the date when the contractor received the certificate, and that, consequently, the notice was served in time.

9.16.3 It was held

(1) The plaintiffs had been entitled to rely upon the defendants' expertise as claims consultants.

(2) Upon the true construction of clause 5.8 of the contract, the final certificate should be issued to the employer and a copy sent to the contractor. If the word 'issue' were to be taken as meaning that the certificate should *reach* the employer, the architect would have to wait until he knew that the employer had received it before sending the copy to the contractor. The interpretation of the clause as propounded by the defendants would, therefore, be absurd.

(3) A certificate was not issued until the architect sent it in the post. Signing it was not sufficient. On the evidence in the instant case, the certificate was posted on 13 October or, at the very latest, 14 October. The last possible date for serving notice, therefore, was on 10 or 11 November. The notice had been given outside the 28-day period, and was thus invalid.

(4) The defendants were found liable to the plaintiffs.

SUMMARY

Claims consultants failed to warn their contractor clients of the need to serve an arbitration notice not later than 28 days after the issue of the final certificate. The contractors lost their entitlements and it was held that they could found an action for negligence against the claims consultants due to their failure to exercise the level of skill expected of an ordinary skilled claims consultant.

9.17   If a delay in the early part of a contract caused by the architect/engineer pushes work carried out later in the contract into a bad weather period causing further delay, can the contractor/subcontractor claim loss and expense resulting from the bad weather delay?

9.17.1   Where delays occur due to bad weather architects and engineers are apt to consider granting extensions of time under the adverse weather clause in the contract. Obligations on the part of the contractor to pay liquidated damages will be avoided but there will be no entitlement to the recovery of additional costs which result from the delay.

9.17.2   Contractors with long experience of submitting claims will usually ask themselves whether carrying out work in the bad weather was due to the knock-on effect of an earlier delay caused by the architect/engineer. In other words whether an earlier delay which was due to some act or omission on the part of the architect/engineer was the reason why in a later working period the contractor was adversely affected by weather and so, had there been no earlier delay, the weather would not have delayed the contractor.

This all goes back to cause and effect. There must be a link between the cause and its effect and the important question is whether such a link can be established between the delay caused by the architect/engineer and the result of the contractor being later delayed through working in bad weather conditions.

9.17.3   In the Canadian case of *Ellis-Donn* v. *The Parking Authority of Toronto* (1978) the plaintiffs were building contractors who successfully tendered to construct a new car park in Toronto. The contract period was 52 weeks but completion was delayed for a period of 32 weeks. One of the causes of delay was a failure by the employer to obtain an excavation permit which delayed the start of the job. The initial period of delay was some seven weeks. It was claimed by the contractor that the initial delay had a knock-on effect which further extended the delay to $17\frac{1}{2}$ weeks made up as follows:

Delay in obtaining the excavation permit	7 weeks
Consequent delay in commencing excavation	$1\frac{1}{2}$ weeks
Consequent delay in obtaining crane	6 weeks
Consequent delay due to extension of work into a winter period	$\underline{3\text{ weeks}}$
	$17\frac{1}{2}$ weeks

9.17.4   The judge, in finding in favour of the contractor, was influenced by the decision in *Koufos* v. *Czarnikow* (1967). It was held in this case that a party who breaches a contract is responsible for the damages that flow from that breach if at the time the contract was entered into the parties to the contract considered there was a real danger or a serious possibility that such breach would give rise to damages.

In giving judgment in the *Ellis-Don* case, the judge held:

In my view the parties to the contract at the time the contract was entered into contemplated or should have contemplated that if the defendant did not have available the necessary excavation permit until seven weeks after the plaintiff had need of it that the plaintiff would be delayed in carrying out its work, and that such delay would throw the subcontractors off schedule which would entail further delays and that such delays would cause the plaintiff damages of the very type or nature it suffered.

9.17.5 Contractors and subcontractors who find themselves delayed by bad weather may be entitled to recover the additional costs as well as time if they are able to show that had it not been for an earlier delay caused by the architect or engineer they would not have been affected by the weather.

SUMMARY

If the contractor is to succeed in claiming loss and expense due to inclement weather on the basis that, but for an earlier delay which was the responsibility of the employer, engineer or architect the bad weather delay would not have arisen, he must be able to link the earlier delay with the effect on later work caused by bad weather.

If there is no direct link or the link in some way is broken then the claim will fail but in appropriate cases the claim for loss and expense as well as time will succeed.

9.18 Who is responsible for the additional costs and delay resulting from unforeseen bad ground conditions – the employer or contractor/subcontractor?

9.18.1 If disputes are to be avoided the contract should make it clear who is responsible for the cost in terms of time and money resulting from encountering unforeseen bad ground. In general terms if the contract is silent on the matter and there is no provision for remeasurement of work on completion the contractor will be deemed to have taken the risk.

9.18.2 The current trend is for employers to seek to pass the risk of this type of matter to the contractor who in turn tries to offload it onto a subcontractor. Usually this is achieved by way of specially drafted conditions of contract. However the majority of the existing standard forms of engineering contract place the risk onto the employer.

Clause 12 of the ICE 6th and 7th Editions entitles the contractor to claim for any additional time and cost which results from:

'physical conditions (other than weather conditions or conditions due to weather conditions) or artificial obstructions which conditions or obstructions could not in his opinion have been foreseen by an experienced contractor.'

Condition 7(3) of GC/ Works/1 1998 Edition includes a similar provision.

9.18.3 The Engineering and Construction Contract (NEC) under clause 60.1 provides for encountering physical conditions to be a compensation event. To qualify these have to be physical conditions which

'● are within the Site,
● are not weather conditions and
● which an experienced contractor would have judged at the contract date to have such a small chance of occurring that it would have been unreasonable for him to have allowed them.'

9.18.4 The FIDIC 1999 Red Book under clause 4.12 entitles the contractor to recover any additional costs incurred as a result of 'physical conditions which he considers to have been unforeseeable'.

Physical conditions are defined by the conditions of contract as:

'natural physical conditions and man made and other physical obstructions and pollutants, which the contractor encounters on the site when executing the works including hydrological conditions but excluding climatic conditions.'

The FIDIC 1999 Silver Book for Turnkey projects takes a different line and under clause 4.12 makes it clear that the contractor is to take the risk and include in his price for 'any unforeseen difficulties or costs, except as otherwise stated in the contract.'

9.18.5 The JCT Forms are silent on the matter – probably because most of the work will be above ground. The risk is therefore taken by the contractor. In the case of *Co-Operative Insurance Society* v. *Henry Boot* (2002), see section 2.1.9, the employer provided a subsoil survey at tender stage which proved to be inaccurate and resulted in the contractor incurring additional costs. The contract incorporated the JCT conditions and the contract documents which stated that:

'the contractor is deemed to have inspected and examined the site and its surroundings and to have satisfied himself as to the nature of the site including the ground and subsoil before submitting his tender.'

It was held that the contractor, under this clause, took the risk of encountering bad ground and lost the case.

9.18.6 With regard to adverse conditions which were foreseeable, Max Abrahamson in his book *Engineering Law and the ICE Contracts* cites the case of *CJ Pearce and Co Ltd* v. *Hereford Corporation* (1968) where:

'contractors knew before tender that a sewer at least 100 years old had to be crossed in the course of laying a new sewer. What was described on the map as "the approximate line of the . . . sewer" was shown on a map supplied to tenderers. The witnesses for both parties accepted that the word "approximate" meant that the contractor would realise that the line of the old sewer might be 10 feet to 15 feet one side or the other of the line shown. The old sewer fractured when the contractors disturbed the surrounding soil within this area.

Held: that the condition could have been "reasonably foreseen", so that even if they had served the necessary notice they would not have been entitled to extra payment under this clause (clause 12) for renewing the old sewer, backfilling the excavation, backheading, etc.'

9.18.7 The case of *Humber Oil Terminals Trustee Ltd* v. *Harbour & General Works (Stevin) Ltd* (1991) appears to have extended a contractors' facility for levying claims under clause 12. A contract was entered into for the construction of three mooring dolphins and the reconstruction of a damaged bathing dolphin at the Immingham Oil Terminal, Humberside. For the purposes of carrying out the work the contractor used a jack-up barge equipped with a 300 tonne fixed crane. The crane was lifting and slewing a large concrete soffit in order to place it on piles which had already been prepared when the barge lifted, became unstable and collapsed. Considerable loss and damage resulted.

The contractor claimed that the collapse of the barge was due to encountering adverse physical conditions which could not reasonably have been foreseen by an experienced contractor and would therefore give an entitlement to levy a claim. It was argued on behalf of the employers that no legitimate grounds for a claim existed. They contended that clause 8(2) applied, under which the contractor was required to take full responsibility for the adequacy, stability and safety of all site operation and methods of construction. The problems arose, the employers said, due to a failure on the part of the contractor to comply with this requirement.

9.18.8 The arbitrator, in deciding whose argument was correct, had to establish the reason for the collapse of the barge. He came to the conclusion that the collapse was caused by the barge moving from an initial small settlement of 5 cm to a substantially large settlement of perhaps 20 cm or 30 cm, at which point the stability of the barge was such that further and progressive collapse occurred.

He decided that the settlement was caused by adverse physical conditions which could not have been foreseen by an experienced contractor, and hence gave rise to a claim under clause 12. On appeal by the employer, the court held that the arbitrator's award was correct.

SUMMARY

Which party to a contract is responsible for unforeseen bad ground should be made clear by the express terms of the contract. The ICE 6th and 7th Editions GC/Works/1 1998, the Engineering and Construction Contract (MEC) and the FIDIC Red Book are all examples of standard forms of contract which place the risk of unforeseen bad ground conditions onto the employer.

If the contract is silent on the matter and there is no provision for remeasurement the contractor will normally be deemed to have taken the risk. This is particularly relevant in lump sum design and construct forms of procurement.

Where employers do not wish to be responsible for information included in subsoil surveys sent with tender enquiry documents, they should make it absolutely clear. Some sort of health warning on the front of the document to the effect that the employer has no liability for the information and the contractor uses it at his peril would seem to fit the bill. The contractor can then carry out the necessary investigations himself, if time permits; if not, he should add a sum to the price to reflect the risk.

9.19 Where one party to a contract is in breach and the injured party incurs loss, what obligations are there on the injured party to mitigate the loss?

9.19.1   Where a breach has occurred, there is a well established legal obligation placed upon the injured party, if the full extent of the loss is to be recovered, to mitigate the loss. In the case of *British Westinghouse* v. *Underground Electric Railways* (1912) Viscount Haldane LC explained the obligation in the following words:

> '... imposes on a plaintiff the duty of taking all reasonable steps to mitigate the loss consequent on the breach, and debars him from claiming any part of the damage which is due to his neglect to take such steps.'

9.19.2   The question of mitigation often arises when a breach of contract occurs resulting in loss or damage. It also can apply where a tort is committed. As can be seen from the *British Westinghouse* case, the plaintiff must take all reasonable steps to mitigate the loss which results from the breach. He cannot recover any loss which he could have avoided. For example, if an employer fails to give proper access to the site at the appropriate time, the contractor would be expected to take plant off hire if it were reasonable and practicable. In other words, the plaintiff cannot recover avoidable loss.

9.19.3   A further matter to be considered is whether the additional costs incurred in taking the reasonable steps are coverable. It was held in *Wilson* v. *United Counties* (1920) that these costs were recoverable. Using the earlier example, the contractor could recover any costs associated with taking the plant off hire. It has been argued that if a contractor is delayed by the employer, he may be obliged to accelerate the works in a reasonable attempt to mitigate the loss. This being the case, under the second rule the acceleration costs should be recoverable. This argument is flawed as the contract normally provides a remedy for employer delays such as an extension of time and payment of loss and expense. However, it could be a different story if the contract such as a JCT form provides a time-scale within which the architect must make a decision concerning an extension of time application, and the architect fails to make a timeous decision.

SUMMARY

Where a breach of contract occurs, the injured party must take all reasonable steps to keep the losses down to a minimum. The losses which could have been avoided by taking reasonable steps will not be recoverable.

9.20 What is meant in legal terms by the words 'consequential loss'?

9.20.1   The words 'consequential loss' frequently appear in construction contracts. Liability for consequential loss is often expressed as being included or excluded. It is tempting to interpret the wording in a similar way to the *Oxford English Dictionary*, being 'events that follow another'.

9.20.2 The decisions in *British Sugar* v. *NEI Power Projects* (1997) and *Simkins Partnership* v. *Reeves Lund & Co Ltd* (2003) dealt with the meaning of consequential loss. The Court of Appeal in British Sugar considered that the proper way to examine this question was by reference to *Hadley* v. *Baxendale* (1854). In this case the court categorised damages under two headings commonly referred to as the first leg and second leg. The first leg deals with losses arising naturally, i.e. according to the usual course of things. By way of contrast, the second leg covers the type of losses which both parties had in mind when the contract was entered into, but which were particular to the contract, sometimes referred to as special damage. It was held that consequential loss is akin to special damage.

9.20.3 The earlier decisions in *Mondel* v. *Steel* (1841) and *Croudace* v. *Cawood's Concrete* (1978) dealt with the meaning of consequential loss in a similar manner.

9.20.4 It would seem the expression that hope dies hard in the human heart applies to some litigants and their legal advisers. In the case of *Hotel Services Ltd* v. *Hilton International Hotels* (2000), the Court of Appeal was again involved in a dispute concerning the meaning of consequential loss. The parties entered into a contract under which the defendants agreed to supply the claimants with Robobars for their hotels.

A Robobar is a hotel minibar which automatically records any removal of its contents and at the same time electronically registers the item concerned on the account of the guest. Robobars were installed in the claimant's hotels on fixed 10-year agreements. They malfunctioned and following attempts to make them work properly, which failed, they were removed. The claimant commenced an action for breach of the implied term of merchantable quality. The judge found in favour of the claimant and awarded damages for overpaid rental, cost of removal and storage and loss of profit.

9.20.5 An appeal was lodged by the defendant on the basis that the terms of the contract included a clause which excluded the liability of the defendant from indirect or consequential losses. It was argued that the judge's award included losses which fell within this category. The Court of Appeal disagreed, as a distinction should be made between normal losses and consequential losses. Normal losses are those which every claimant in a like situation will suffer, whereas consequential losses are those peculiar to the particular claimant's circumstances and not foreseeable.

On this view consequential losses are anything beyond the normal measure of damages. The losses included in the judge's award were all normal losses and fell outside the term consequential losses.

SUMMARY

Consequential loss does not include those which are normal and usual but comprises losses peculiar to the particular circumstances and not foreseeable.

9.21   **Is it possible to include in a contract a daily or weekly rate which will be paid to the contractor in respect of loss and expense or additional cost resulting from delays caused by the employer?**

9.21.1   Where, due to delays on the part of the employer, the completion date is affected and the contractor incurs direct loss and expense or additional costs, most standard forms of contract provide for the contractor to recover financially. The relevant clauses usually require the contractor to submit a written notice and supporting detail. The architect, engineer or quantity surveyor is normally empowered to ascertain the contractor's financial entitlement which is then certified and paid.

9.21.2   This process can be time consuming and expensive with payment to the contractor often the subject of a long delay. It has been suggested that there should be provision in the contract for the inclusion of an all-inclusive daily or weekly sum which will be paid to the contractor in the event of delays to completion being caused by the employer. The daily or weekly rate could be included in the contractor's tender submission or be the subject of agreement between the parties post tender submission but prior to the contract being entered into.

9.21.3   The main disadvantage of this arrangement is that it only deals with the financial effects of a delay to the completion, leaving disruption and delays which only affect progress but not completion to be dealt with separately. Contractors will need to appreciate that the sum will also include for any legitimate claims which may be received from subcontractors. Finally, during pre-contract negotiations contractors may well find themselves under pressure to agree to an unrealistically low daily or weekly amount.

SUMMARY

It is possible to include in a contract a daily or weekly rate which will be paid to the contractor in respect of loss and expense or additional costs resulting from delays caused by the employer. The daily or weekly rate will include for any claims received from subcontractors. Claims for disruption and delays which only affect the progress but not completion will not be included and will therefore be the subject of a separate submission.

10

PRACTICAL COMPLETION AND DEFECTS

10.1 How are practical completion and substantial completion under the JCT and ICE conditions defined?

10.1.1 A point of debate on building contracts frequently relates to practical completion. The issue of a certificate of practical completion by the architect will usually lead to a sigh of relief from the contractor. Any liability for liquidated damages ceases; the employer becomes obliged to insure; retention is released and the defects period begins to run. It is, therefore, not surprising that architects and contractors frequently argue as to whether or not practical completion has been achieved.

10.1.2 Where a dispute occurs, a first port of call is usually the conditions of contract. JCT 98 does not unfortunately define practical completion. Clause 17.1 however provides for the following:

> 'When in the opinion of the Architect Practical Completion of the Works is achieved ... he shall forthwith issue a certificate to that effect and Practical Completion of the Works shall be deemed for all the purposes of this Contract to have taken place on the day named in such certificate.'

10.1.3 The question of practical completion under a JCT form of contract has been the subject of referral to the courts on more than one occasion. Unfortunately there is still no precise definition.

10.1.4 In *J. Jarvis & Sons* v. *Westminster Corporation* (1978) the House of Lords had to decide whether the main contractor was entitled to an extension of time under JCT 63 where delays occurred due to remedial works undertaken by the nominated piling subcontractor after the piling work was completed. Practical completion of the subcontract works became relevant as the court held that no extension of time was due in respect of delays caused by remedial works to the piling after the piling had been completed. Lord Justice Salmon started the ball rolling with this version of what is meant by practical completion:

> I take these words [practical completion] to mean completion for all practical purposes, that is to say for the purposes of allowing the employer to take possession of the works and use them as intended. If completion in clause 21 meant completion down to the last detail, however trivial and unimportant, then clause 22 would be a penalty clause and as such unenforceable.

Contractors no doubt were more than pleased with this definition but unfortunately the waters were muddied with the definition given in the same case by Lord Dilhorne who said in a dissenting judgment:

> The contract does not define what is meant by practical completion. One would normally say that a task was practically completed when it was almost but not entirely finished, but practical completion suggests that that is not the intended meaning and what is meant is the completion of all the construction that has to be done.

10.1.5 The courts made a further attempt at a definition in *H.W. Nevill (Sunblest) Ltd v. William Press & Son Ltd* (1981). In this case defects occurred after practical completion of a preliminary works contract which delayed a follow-on contract. Again what constitutes practical completion was relevant. It would appear the judge in this case sided with the views of Lord Dilhorne in saying:

> I think the word 'practically' in clause 15(1) gave the architect a discretion to certify that William Press had fulfilled its obligation under clause 21(1), where very minor de minimis work had not been carried out, but that if there were any patent defects in what William Press had done the architect could not have given a certificate of practical completion.

10.1.6 A more recent case where practical completion was at issue is *Emson Eastern Ltd* v. *EME Developments Ltd* (1991). Emson were the contractors and EME developers for the erection of business units. JCT 80 formed the basis of the contract.

Judge John Newey QC, in arriving at the meaning of completion of the works, took account of what happens on building sites. He considered that he should keep in mind that building construction is not like the manufacture of goods in a factory. The size of the project, site conditions, use of many materials and employment of various types of operatives make it virtually impossible to achieve the same degree of perfection as can a manufacturer. His view was that it must be rare for a new building to have every screw and every brush of paint correct. Further a building can seldom be built precisely as required by the drawings and specification. Judge Newey in considering the meaning of practical completion thought he stood somewhere between Lord Salmon and Viscount Dilhorne in the Jarvis case.

10.1.7 The Court of Appeal of Hong Kong in *Big Island Contracting (HK) Ltd* v. *Skink Ltd* (1990) had to decide whether practical completion had been achieved.

The plaintiffs were contractors for work on the 12th and 13th floors of the defendant's building under a contract which provided payment of 25% of the price upon practical completion. The defendants went into occupation of the building. The plaintiffs contended that the works had been practically completed and issued proceedings seeking for 25% of the agreed price, i.e. nearly HK$ 100 000.

District Judge Yam found that practical completion had not been achieved since there were defects which would have cost between HK$ 40 000 to HK$ 60 000 to rectify and that the plaintiffs had failed to execute an important part of the modification of the sprinkler system on the 13th floor for which a provisional sum of HK$20 000 had been allowed. The judge found that this defect affected the safety of the system and would take between two and ten days to correct. He gave judgment for the defendants. The plaintiffs' appeal against the decision was dismissed.

Occupation by the employer does not in itself, therefore, constitute practical completion.

10.1.8 A comprehensive definition of 'substantial completion' as it applies to the ICE conditions appears in *Engineering Law and the ICE Contracts*, 4th edition by Max W. Abrahamson at page 160, as follows:

> 'The *Concise Oxford Dictionary* equates "substantial" with "virtual" which is defined as "that is such for practical purposes though not in name or according to strict definition". It is at least clear on the one hand that the fact that the works are or are capable of being used by the Employer does not automatically mean that they are substantially complete ("any substantial part of the Works which has both been completed ... and occupied or used") and on the other hand that the Engineer may not postpone his certificate under this clause until the works are absolutely completed and free of all defects. The many reported cases on the question of "substantial" completion in relation to payment under an entire contract, a different legal problem, are of doubtful relevance. Obviously both the nature and extent of the uncompleted work or defects are relevant, and to say that substantial completion allows for minor deficiencies that can be readily remedied and which do not impair the structure as a whole is probably an accurate summary of what is a question of fact in each case.'

10.1.9 The difference between JCT 98 and the ICE 6th and 7th Editions is that under the latter the engineer may issue a certificate of substantial completion with outstanding work still to be done provided the contractor gives an undertaking to complete the outstanding work in the maintenance period.

SUMMARY

There seems to be little difference in the meaning of practical completion and substantial completion. From the various authorities available on this subject, Judge Newey in *Emson Eastern* v. *EME* seems the most sensible. He advised architects that, when issuing a certificate of practical completion, they should bear in mind that construction work is not like manufacturing goods in a factory. His view was that it must be rare for a new building to have every screw and every brush of paint correct.

10.2 Where an employer takes possession of a building or engineering facility before all the work has been completed, can the contractor rightfully claim that practical completion or substantial completion has been achieved?

10.2.1 Some contracts specifically state that, where a building or facility is completed to the stage where it is fit for occupation and by the employer, then practical completion has been achieved. The RIAI (2002) Edition used in the

Republic of Ireland under clause 31 states that practical completion means
'the works have been carried to such a stage that they can be taken over and
used by the Employer for their intended purpose...'.

Other contracts such as those published by the JCT merely make reference
to the architect being obliged to issue a certificate of practical completion
when the works have reached the stage of being practically completed.

10.2.2 Practical completion of the works is a defining moment on most construction
contracts. Retention becomes due for release, liability for damage passes to
the employer and of course any obligation on the contractor's part to pay
liquidated damages ceases. Most standard forms of contract include both a
clause for practical completion of the whole of the works and also for the
employer taking possession of part of the works. The wording normally
states that if the employer during a period of overrun takes possession of
part of the works before practical completion of the whole of the works, then
the contractor's obligation to pay liquidated damages will be at a reduced
level.

10.2.3 The decision in the case of *Skanska Corporation* v. *Anglo-Amsterdam Corporation* (2002) shows that complications can arise where part possession by the
employer occurs. Skanska Corporation undertook to construct an office
facility in Edinburgh for the Anglo-Amsterdam Corporation employing a
JCT 81 With Contractor's Design standard form of contract. Clause 16 of the
contract deals with practical completion and the process requires the
employer's agent to provide a written statement to indicate when it has been
achieved. The standard wording had been amended to read that the statement will only be issued when the employer's agent is satisfied that any
unfinished work is very minimal and of a minor nature. A dispute arose
between the parties as to the date by which the contractor had achieved
practical completion. The matter was referred to an arbitrator who had to
decide whether practical completion took place on 12 February 1996 or 25
April 1996.

10.2.4 It seems that Anglo-Amsterdam had leased the premises to ICL who were
anxious to gain access to enable the fit-out to commence. By the time 12
February 1996 had arrived, work was still incomplete. The main problem
was the air-conditioning, which was not fully functioning, and Skanska's
failure to produce operating and maintenance manuals. The seasoned
observer may consider that this is not unusual. The situation, however,
should not hold up the fit-out and therefore ICL moved in. Skanska did not
finish off the incomplete items until 25 April 1996 and as a result Anglo-
Amsterdam levied liquidated damages for a failure to complete on time.
Skanska argued that as ICL had moved in on 12 February 1996, liquidated
damages could not be deducted for the period after that date. This called for
a close examination of the wording in the contract. Clause 16 was stringent,
bearing in mind that there was no entitlement to an employer's agent's
statement that practical completion had occurred if work, except for that of a
very minimal and minor nature, was still outstanding. No help here for
Skanska, as the air-conditioning was not working by 12 February 1996. The
wording of clause 17 did not seem to apply either as it only deals with a
situation where the employer takes possession of a part of the works, in

other words partial possession. What appears to have occurred is that ICL took partial possession of the whole of the works. They did not take full possession of the whole of the works as Skanska were engaged in finishing off the work, while ICL were undertaking the fit-out.

10.2.5   Judge Thornton QC took a very mature and practical view of the situation in finding in favour of Skanska. Clause 17, which deals with partial possession, states that if the employer takes over a part of the building then as far as that part of the works is concerned the contractor is deemed to have achieved practical completion. Responsibility for the part taken over rests squarely with the employer for such matters as damage and health and safety. This would apply even if the clause 16 definition of practical completion had not been achieved. Whilst the contract did not deal with the situation where the employer takes possession of the whole of the works before practical completion, the same principle should apply as where the employer takes possession of part of the works. In other words, practical completion is deemed to have taken place. The word 'deemed' is an interesting one in that it means that for all legal purposes an event has happened even if as a fact it has not. In other words, if practical completion is deemed to have occurred it would not matter if the building had no roof on. In this case it was only the air-conditioning which was not completed.

10.2.6   It is not in all cases of employers gaining access that practical completion occurs, which became clear in the case of *Impresa Castelli SpA* v. *Cola Holdings Ltd* (2002). Clause 23.3.2 of the JCT contracts gives the employer an entitlement to gain access to part or all of the works. This presence on the site does not affect his entitlement to deduct liquidated damages. Cola, the employer, engaged Impresa Castelli for the construction of Kingsley Hall Hotel in Great Queen Street, London. Work was running late, an obligation to pay liquidated damages arose and Cola levied a claim of £1.2m.

Impresa Castelli argued that Cola had taken partial possession of the whole of the works under clause 17.1, and in keeping with the *Skanska* decision no liquidated damages were payable. Judge Thornton, who heard the *Skanska* case, did not agree. Based upon the facts, he considered that occupation by Cola did not amount to taking partial possession as envisaged by clause 17.1. The situation led the judge to believe that Cola had merely gained access to the site as envisaged by clause 23.3.2 and therefore liquidated damages were payable.

10.2.7   The ICE 7th Edition, under clause 48(4), provides for the engineer to issue a certificate of substantial completion in respect of parts of the work. Access to the site is provided for under clause 37. Whilst the wording in the ICE conditions is not as detailed as the JCT, it is considered that if the ICE 7th Edition conditions had applied in the *Skanska* and *Impresa Castelli* cases, the decisions of the court would have been no different.

SUMMARY

It would make the situation clear if contracts state that should the employer at any time before practical completion of the works take possession of a part of or the whole of the works, then practical completion of the part or the whole of the works taken into possession is deemed to have taken place.

In the absence of such wording but in the light of the decision in *Skanska Corporation* v. *Anglo-Amsterdam Corporation* (2002), it is likely that, if an employer takes possession of a building or facility, practical completion will be deemed to have occurred. This, however, may be confused with access being granted to the employer where practical completion is not deemed to have occurred. There is always likely to be scope for argument as to whether an employer has been granted occupation when practical completion will be deemed to have occurred, to be contrasted with the employer gaining access only, which does not involve practical completion.

10.3   When does practical completion occur under a subcontract where the DOM/1 conditions apply?

10.3.1   It is important for contractors or subcontractors to establish with some precision the date when practical completion of the subcontract has been achieved. Uncertainty may affect the contractors' right to claim for late completion against the subcontractor. Liability for damage to the subcontract works and release of retention are also affected.

10.3.2   In the case of *Vascroft (Contractors) Ltd* v. *Seeboard plc* (1996) a dispute arose as to when practical completion of a subcontract works occurred. The conditions of the subcontract were DOM/2 and the dispute was referred to arbitration in accordance with the provision of the contract.

Under DOM/2 clause 14.1 the subcontractor is required to notify the contractor in writing of the date when, in his opinion, the subcontract works are practically completed. The arbitrator found in this case that, even if the subcontractor gave no such notice, practical completion was still not deemed to have occurred as set out in clause 14.2. This clause states that if the contractor dissents in writing, practical completion occurs on such date as is agreed and in the absence of agreement, practical completion is deemed to occur on the date of practical completion of the main contract works as certified by the architect.

Vascroft appealed against this finding.

The judge said:

> In my judgment, where, as here, the parties have entered into a standard form of contract (with some amendments to bring it into line with the nature of the work) it is necessary to look at the whole of the contract. It is quite clear that there are many provisions of this contract which depend for their effective operation on a firm date for practical completion being established under clause 14, either by operation of the mechanism there provided or by agreement.
>
> The machinery of clause 14 starts with a notice being given by the subcontractor. In my judgment, the words 'The subcontractor shall notify the contractor in writing of the date when, in his opinion, the subcontract works are practically completed' impose an obligation on the subcontractor to do so. The contractor will undoubtedly have his own view as to whether or not the subcontract works are complete, but this contract does not require him to express it unless and until he

receives a notice from the subcontractor. If the subcontractor had given a notice, but the contractor duly dissented from it and no agreement could be reached, then the deeming provisions would establish a firm date, i.e. the date of practical completion of the main contract works. I do not consider it right that, if a subcontractor is in breach of his obligation to serve written notice, the date for practical completion fails to be established as a matter of fact. If this interpretation were correct, it could, in many instances, suit a subcontractor not to give any notice at all rather than a notice which he knew would not be accepted.

The judge concluded that in the absence of a written notice from the subcontractor the situation with regard to practical completion would be the same as would occur if the subcontractor served notice which was dissented from by the contractor and no agreement was reached. In other words in the absence of notice practical completion of the subcontract will occur on the date of practical completion of the main contract as certified by the main contractor.

SUMMARY

Clause 14.1 of DOM/1 and DOM/2 requires the subcontractor to serve notice on the main contractor when in his opinion the subcontract works are practically completed. If the main contractor dissents in writing practical completion occurs on such date as is agreed. In the absence of agreement practical completion is deemed to occur on the date of practical completion certified by the architect under the main contract.

If the subcontractor fails to serve notice then again practical completion of the subcontract work will be deemed to occur on the same day as practical completion certified by the architect under the main contract.

Subcontractors operating under DOM/1 and DOM/2 are advised to ensure that written notice is sent to the main contractor as soon as the subcontractor considers his work has been completed.

10.4 Where at the end of the defects liability/maintenance period the architect/engineer draws up a defects list but due to an oversight omits certain defects, and a second list is prepared after the defects on the first list have been completed, will the contractor/subcontractor be obliged to make them good?

10.4.1 To prevent stale claims being levied for breach of contract, the Limitation Acts lay down periods of time within which actions must be commenced. These periods of time seem generous.

The Limitation Act 1980 states by section 7:

'An action to enforce an award, where the submission is not by an instrument under seal, shall not be brought after the expiration of six years from the date on which the cause of action accrued.'

and by section 8

'An action upon a speciality shall not be brought after the expiry of twelve years from the date on which the cause of action accrued.'

This means in reality that if the contract is a verbal or written contract an action will be time barred if not commenced within six years of practical completion. Where the contract is a speciality, in other words under seal or expressed as a deed, the period is twelve years.

Where a construction contract includes a defects or maintenance period which is less than the timescales laid down in the Limitation Acts do the provisions of the defect or maintenance clauses shorten the effect of the Limitation Acts?

10.4.2   JCT 98 and other JCT forms, ICE 6th and 7th Editions and GC/Works 1/ 1998 Edition all include a defects liability or maintenance period. The purpose of these defects or maintenance periods is to allow the contractor an opportunity of making good his own defects. Whilst it may not be obvious to many contractors and subcontractors, these clauses actually bestow a benefit upon them. In this context *Keating on Building Contracts* 5th edition at page 247 states:

'The contractor's liability in damages is not removed by the existence of a defects clause except by clear words, so that in the absence of such clear words the clause confers an additional right and does not operate to exclude the contractor's liability for breach of contract ... But it is thought that most defects liability clauses will be construed to give the contractor the right as well as to impose the obligation to remedy defects which come within this clause.'

In other words defects in construction work amount to a breach of contract entitling the employer to claim damages: *HW Nevill (Sunblest)* v. *William Press and Son* (1981). In the absence of a defects clause and where defects in the work appear after practical completion the employer would be within his rights to employ others to make good the defects and to charge the contractor with the cost. The presence of a defects clause gives the contractor the right to remedy his own defects, the cost of which should be less than would be the case if others carried out the work.

Hudson's *Building and Engineering Contracts* 11th edition at paragraph 5.050 says:

'Since such work can be carried out much more cheaply and possibly more efficiently by the original contractor than by some outside contractor brought in by the building owner, defects clauses in practice confer substantial advantages on both parties to the contract.'

10.4.3   JCT 98 under clause 17.2 requires the architect to prepare a schedule of defects not later than 14 days after the end of the defects liability period. No specific reference is made in either the ICE 6th or 7th Edition or GC/Works/ 1 1998 to a defects list but it is nonetheless common practice.

10.4.4   A question often raised is whether a contractor is obliged to make good defects if under a JCT 98 contract the architect produces the defects schedule outside the 14 day period or alternatively if, having issued a schedule with which the contractor has complied, he produces a second schedule which lists more defects.

When answering the question the comments made earlier should be borne in mind. Defects in the contractor's works amount to breaches of contract for which the employer is entitled to damages and this is not excluded by the defects clause. All the defects clause does is to give the contractor the right to make good defects. The contractor would be well advised to make good defects on late, second and subsequent lists if these situations were to occur.

SUMMARY

It would seem that a failure by the architect/engineer to issue a defects schedule on time or the issue of a second one would not amount to a waiver of the employer's rights. The contractor may be able to make out a case that a payment should be made for the costs of making two visits as a result of the architect's failure to include all defects on the original list. It is, however, advisable for him to take the cheaper option of exercising his right to remedy his own defects rather than have the employer call in an outsider and sue for the costs.

The defects or maintenance clauses do not overrule the provisions of the Limitation Acts.

10.5 Is a contractor/subcontractor absolved from any liability if the employer refuses him access to make good defects because he chooses to make them good himself?

10.5.1 The standard forms of contract provide for contractors to make good defects following practical or substantial completion during the defects period.

Clause 17.3 of JCT 98 provides for the architect to give instructions to the contractor that certain or all defects are not to be made good. This being the case the clause goes on to provide that an appropriate deduction is to be made from the contract sum. No assistance is given in the clause as to how the amount of deduction is to be calculated.

Most of the other standard forms make no provision for making good defects being omitted.

10.5.2 Problems can arise where employers make their own arrangements to make good defects.

The decision in *William Tomkinson & Sons Ltd* v. *The Parochial Church Council of St Michael and Others* (1990) aptly deals with this point.

William Tomkinson, the contractor, was employed by the church council to carry out restoration works at a parish church in Liverpool. The contract was let using the JCT Minor Works 80.

A dispute arose concerning certain defects and damage to the works including damage to plasterwork caused by hammering, damage to woodblock flooring, plasterwork, rooflights and their flashings etc. plus many more. The contractor's defence was that either the damage was not of his making or alternatively he was protected by the provisions of clause 2.5 of the conditions of contract.

This clause makes the contractor liable for any defects, excessive shrink-age or other faults which appear within three months of the date of practical completion due to materials or workmanship not being in accordance with the contract.

The defects and damage which formed the basis of the church council's case had been discovered and remedied by other contractors on instruction of the church council prior to practical completion.

Both parties agreed that, for clause 2.5 to be effective, notice of the defects must be given to the contractor. It was also common ground that some but not all items of defect or damage resulted from defective workmanship by the contractor.

No warning was given to the contractor prior to the church council instructing other contractors to carry out the work.

It was argued on behalf of the contractor that the church council, in arranging for having defective work corrected by others, prevented the contractor from exercising his rights to correct defects and therefore the contractor had no liability.

10.5.3   The court in arriving at a decision was influenced by the wording in *Hudson's Building and Engineering Contracts* 10th edition at pages 394–7:

> 'It is important to understand the precise nature of the maintenance or defects obligations. It is quite different from the Employer's right to damages for defective work, under which he will be able to recover the financial cost of putting right work either by himself or another contractor. Since maintenance (defects) can usually be carried out much more cheaply by the original contractor than some outside contractor ... so the contractor not only has the obligation but also in most cases it is submitted the right to make good as its own cost any defects.'

The court went on to consider that workmanship which falls short of the standard required by the contract and which the employer remedies prior to practical completion still constitutes a breach of contract.

It was held, in finding in favour of the church council, that their entitle-ment was to recover damages from the contractor subject to proof that these were attributable to workmanship or materials which fell below the contractual standard. The amount of damages which the church council would be entitled to recover was not, however, their outlay in remedying the damage, but the cost which the contractors would have incurred in reme-dying it if they had been required to do so – a sum anticipated to be much less than the actual remedial costs.

The answer to the question therefore is that a contractor is not absolved from liability if he is refused access but damages recoverable by the employer are limited to the amount it would have cost the contractor had he been given an opportunity to make good the defects.

SUMMARY

JCT 98 under clause 17.3 gives the architect power to instruct the contractor not to make good defects in which case an appropriate deduction will be made from the contract sum. Most of the other standard contracts make no such provision.

If defects are not made good by the contractor but the employer arranges for the work to be carried out by others the employer will only be entitled to recover from the contractor what it would have cost the contractor had he in fact made good the defects. Agreement as to what those costs might have been could be difficult to achieve.

10.6 Most subcontractors provide for the release of the final balance of retention only when the period included in the contract for correcting defects has expired and all defects under the main contract have been made good. If the main contractor or other subcontractors are dilatory in making good defects, is there any mechanism to enable the subcontractor to secure an early release of retention?

10.6.1 Subcontractors regularly complain of the huge amounts of money tied up in retention. The sums usually exceed a year or two's profits. Where a JCT form of contract is employed, subcontractors usually have to wait until the architect has issued the certificate of making good defects under the main contract before they receive the final balance of retention. This will only occur when all defects have been made good to the architect's satisfaction. Other standard forms of contract such as the GC/Works/1 and ICE are similarly worded. Subcontractors are often disadvantaged when they have corrected their own defects but the certificate of making good defects has not been issued due to a failure on the part of the main contractor or other subcontractors.

10.6.2 In the case of *Pitchmastic* v. *Birse Construction* (2000), a dispute arose between the main contractor and a subcontractor concerning the release of retention. The main contract for the construction of a new distribution centre for Tesco in Milton Keynes employed a JCT With Contractor's Design contract. Pitchmastic were subcontractors for the roofing works, the subcontract being an amended DOM/2. The subcontract works were completed according to Pitchmastic on 26 February 1998, followed by practical completion of the main contract works on 22 May 1998. A defects liability period of 12 months was included in the main contract. As at 21 May 1999, Pitchmastic's work was defects free, but defects remained to be corrected under the main contract. The final half of retention in the sum of £33 651 was held in respect of Pitchmastic's work. Pitchmastic commenced an action for the recovery of balances it claimed to be due under the final account including release of the retention.

10.6.3 Birse Construction argued that release of retention was not due until the certificate of making good defects had been issued. In finding in favour of the main contract, Judge Dyson said:

> 'It is not sufficient for Pitchmastic merely to show that the architect withheld the certificate because there were outstanding defects in the work that had been car-

ried out by Birse or other subcontractors. Provided that Birse and its sub-contractors were proceeding with reasonable diligence to make good the defects, Birse was not preventing the issue of the certificate. Some defects take longer to put right than others.'

SUMMARY

A situation may arise where the period for correcting defects has expired. One of the subcontractors has completed making good defects in its work but there is still defects correction required on the part of the main contractor and other subcontractors. It was held in *Pitchmastic* v. *Birse Construction* (2000) that provided the correction of outstanding defects is proceeding with regular diligence, the contractor will not be obliged to release the out-standing retention to the subcontractor.

10.7 Can an employer recover from the contractor the costs involved where it became necessary to employ an external expert to demonstrate that work was defective?

10.7.1 Telling a contractor that the defects liability period provides him or her with a benefit is likely to be an invitation for a rude response. Most contractors consider that an obligation to return to site some six, nine or twelve months after the architect or engineer has issued a certificate signifying that satis-factory completion has been achieved, provides no discernable advantage.

10.7.2 The decision in the Scottish case of *Johnston* v. *W.H. Brown Construction (Dundee) Ltd* (2000) provides otherwise. A contract was let to construct an office block employing the Scottish Building Contract With Contractor's Design. Following practical completion, defects appeared in the external envelope. The employer appointed a firm of architects, HRP, to investigate. Their report identified a number of defects resulting from a failure to follow the Building Regulations. These defects were remedied by the contractor without admitting liability. The employer commenced an action against the contractor to recover the fees paid to the architect, together with solicitor's fees in connection with the provision of contractual advice, and the employer's management costs. In all a sum of around £80 000 was claimed.

10.7.3 The court found against the employer in that clause 16 of the contract provides its own remedy where defects are discovered during the defects period. It is the duty of the employer to produce a defects schedule at the end of the defects period, and the contractor's obligation to put them right. If the contractor carries out the remedial work, the employer has no other remedy.

10.7.4 There is no provision in the contract which allows the employer to recover the costs of identifying the defects, taking advice, instructing the contractor and subsequently inspecting the remedial works. In the absence of such a clause, defects appearing after completion would be treated as a breach of contract, leaving the employer to recover the expen-diture as common law damages.

SUMMARY

Where the standard forms of contract are used an employer is unable to recover the cost of employing external consultants to demonstrate that work was defective. The situation may be different where standard forms are not used, when this type of cost may be recovered as damages for breach of contract.

10.8 Where a dispute arises between employer and contractor which includes defective work carried out by a subcontractor and is the subject of legal proceedings which are settled by the contractor making a payment to the employer, is the subcontractor obliged to reimburse the contractor for the cost of remedying defective work even though he considers the subcontract works contain no defects?

10.8.1 Where a dispute arises which involves employer, contractor and subcontractor, difficulties can arise in getting all three to agree to a compromise arrangement. It is often the case that legal proceedings are commenced involving all three. Pressure is applied for the parties to settle the dispute and it is not uncommon for the contractor and employer to reach a settlement, leaving the contractor to arrive at an accommodation with the subcontractor. The contractor will usually attempt to impose upon the subcontractor that part of the settlement with the employer which relates to the subcontractor's work.

10.8.2 In the case of *Bovis Lend Lease Ltd* v. *RD Fire Protection Ltd* (2003), a dispute arose concerning a claim from the management contractor, Bovis, against RD Fire Protection, a subcontractor who undertook the dry lining and fire protection. Bovis had been appointed as the management contractor on the Braehead Shopping Centre in Glasgow. A major row broke out between Braehead, the developer, and Bovis, with claims flying in both directions. Bovis alleged that they were owed £37m as the balance due on the final account and damages for breach of contract, whilst Braehead counterclaimed in the sum of £65.8m in respect of alleged mismanagement, defective work and overvaluation of the works. Proceedings were commenced and as part of the process Bovis joined RD Fire Protection as a joint defendant on the basis that part of the counterclaim related to the quality of work undertaken by RD Fire Protection. Bovis and Braehead entered into a compromise arrangement whereby Braehead paid Bovis a sum of £15m in full and final settlement of all matters in dispute which included the alleged defective work carried out by RD Fire Protection.

10.8.3 The settlement reached between Braehead and Bovis was on a global basis in that there was no build-up which showed how the amount had been calculated. The cost of correcting the defective work formed part of the counterclaim but it was not clear whether or not the agreed figure included any provision in respect of the alleged defective work. Bovis, however, claimed from RD Fire Protection the full cost of the remedial works as pleaded by

Braehead. They sought to bypass the settlement entirely and proceed as if it had never taken place. This was obviously wrong as Bovis were only entitled to recover whatever loss they could prove resulted from the defective work. It was possible for there to be some provision in the global settlement figure for the defective work; the more likely explanation, however, was that as the settlement figure was on a global basis the sum could not be identified. It was open therefore for Bovis to request the court to ascertain the sum involved, which could then be passed on to RD Fire Protection. Bovis did not use this approach, but instead passed on the total sum originally claimed by Braehead. The court rejected the case brought by Bovis and found for RD Fire Protection.

10.8.4   It is possible to pass down to a subcontractor a part of a settlement agreement reached with an employer if it can be shown that the sum to be passed down is reasonable. This situation was explored in *Biggin* v. *Permanite* (1951), *P&O Developments* v. *The Guys and St Thomas's National Health Trust* (1998) and *Sainsbury* v. *Boardway Malyan Earnest Green* (1998). It was decided in these cases that the question is not whether the plaintiff who entered into the compromise and seeks a contribution acted reasonably, but whether the settlement was a reasonable one.

SUMMARY

When a dispute between an employer and contractor involving defective work carried out by a subcontractor is settled, the contractor cannot as of right recover the amount included in the settlement in respect of the subcontractor's work. It is open to the contractor if the dispute is the subject of legal proceedings, to request the court or arbitrator to identify the sum in a settlement agreement. In the absence of the amount being identified by the court or arbitrator, then the amount which the subcontractor is obliged to pay must be reasonable.

11
RIGHTS AND REMEDIES

11.1 **Where a contractor/subcontractor whose tender is successful receives a letter of intent, is he at risk in commencing work or ordering materials or carrying out design? If the project is abandoned before a contract is signed will there be an entitlement to payment?**

11.1.1 To establish that a contract has been concluded not only requires evidence of agreement by the parties on all the terms they consider essential, but also sufficient certainty in their dealings to satisfy the requirement of completeness. Letters of intent traditionally fail on both counts since they are usually incomplete statements preparatory to a formal contract. A letter of intent as traditionally worded is binding upon neither party: *Turriff Construction Ltd* v. *Regalia Knitting Mills Ltd* (1971).

11.1.2 Fast track construction methods often overtake the procedures for drawing up the contract, which in many instances lack the necessary urgency. This has resulted in increasing use of letters of intent for purposes other than the original one, which was little more than to inform the contractor/subcontractor that his tender was successful and that a contract would be entered into at some stage in the future.

This original function has changed in recent times. It is now common practice to include in a letter of intent an instruction to commence design, order materials, fabricate and even start construction on site in anticipation of the contract being entered into.

11.1.3 Arguments arise in particular cases as to whether the letter of intent itself constituted a contract and, if not, whether the negotiations which followed result in a concluded contract. If it is held that there was never a contract entered into, further disputes arise as to the basis on which payment is due for the work carried out in accordance with instructions contained in the letter of intent.

11.1.4 In *British Steel Corporation* v. *Cleveland Bridge Ltd* (1984) the court had to deal with the question as to whether a particular letter of intent created a contract. In the context of the case Lord Justice Goff said:

> Now the question is whether, in a case such as the present one, any contract has come into existence must depend on a true construction of the relevant commu-

nications which have passed between the parties and the effect (if any) of their action pursuant to those communications. There can be no hard and fast answer to the question whether a letter of intent will give rise to a binding agreement; everything must depend on the circumstances of the particular case.

The judge went on to say that if work is done pursuant to a request contained in a letter of intent it will not matter whether a contract did or did not come into existence because, if the party who has acted on the request is simply claiming payment, his claim will usually be based on a quantum meruit. Unfortunately, this seems a rather simplistic view as there is no hard and fast rule as to what constitutes a quantum meruit payment.

It is of little advantage to a contractor or subcontractor to learn that he is entitled to a payment if there is no agreement as to how much the payment will be or how it will be calculated.

11.1.5 In the case of *Kitsons Insulation Contractors Ltd* v. *Balfour Beatty Buildings Ltd* (1989) the court had to decide whether a letter of intent sent by the main contractor to a subcontractor created a contract.

Balfour Beatty was appointed main contractor for Phase 1 of the White City Development for the BBC. Kitsons submitted a tender to Balfour Beatty in October 1987 in the sum of £1.1 m for the design, manufacture, supply and installation of modular toilet units and accessories. In the period which followed a large number of variations were made by Balfour Beatty as to the details of the work required by them, and as a result Kitsons revised their tender upwards by £70 000.

Balfour Beatty sent a letter of intent to Kitsons in March 1988. The general gist of the letter was that Balfour Beatty intended to enter into a subcontract with Kitsons using the standard subcontract DOM/2 1981 edition amended to suit Balfour Beatty's particular requirements which was to be forwarded in due course. The subcontract amount was to be £1 162 451 less 2½% discount on a fixed price lump sum basis. Finally, the letter of intent requested Kitsons to accept the letter as authority to proceed with the subcontract works.

Kitsons, as requested, signed and returned the letter as acknowledgement of receipt and then commenced work.

It was not until August 1988 that Balfour Beatty drew up and submitted a formal subcontract to Kitsons. Accompanying the subcontract was a letter indicating an acceptance of Kitsons' offer. The letter went on to say that payment was not to be made until the subcontract had been signed by Kitsons and returned.

Kitsons did not sign and return the subcontract. Their stated reasons were twofold: firstly, the subcontract received for signing from Balfour Beatty included a number of variations not provided for in the price and secondly, the main item of cost related to off-site fabrication and, to safeguard their cashflow, Kitsons had included with their tender an activity schedule for interim valuations and payments. No provision had been made in the subcontract for including this schedule or the cost of off-site fabrication.

Following commencement of the work by Kitsons payments totalling £992 767 were made by Balfour Beatty based upon the subcontract conditions DOM/2 with amendments which Balfour Beatty considered applied.

Kitsons claimed that no binding subcontract had been concluded by the parties and claimed to be entitled to payment on a quantum meruit basis – in other words, a reasonable sum for the work. It was Kitsons' opinion that the amount paid by Balfour Beatty fell £660 000 short of what constituted a reasonable amount, i.e. about £500 000 more than the contract sum.

Kitsons commenced an action against Balfour Beatty and the court had to decide a preliminary point as to whether Balfour Beatty in sending the letter of intent to Kitsons created a contract. It was held that no contract had been concluded as the parties had not arrived at the stage where it could be said that full agreement had been reached between them – the matters outstanding, in particular the method of payment, were too significant for a contract to have come into place.

Whilst this decision settled that a contract had not come into being, it left unanswered how much Kitsons were entitled to be paid on the quantum meruit basis – a matter for negotiation by the parties which, in the event, did not have to come back before the court.

11.1.6 In *Mitsui Babcock Engineering Ltd* v. *John Brown Engineering Ltd* (1996) a dispute arose as to whether a contract had come into being. In October 1992 Mitsui Babcock Engineering Ltd sent a proposed form of contract to John Brown Engineering Ltd which was subject to the MF/1 Conditions. Clause 35 of the conditions which was headed 'performance tests' was struck out and a marginal note 'to be discussed and agreed' written in. The document was signed by both parties but there was no agreement concerning performance tests and liquidated damages for failure to achieve the performance tests. Despite the parties' inability to reach agreement on these matters the court held that there was none the less a binding contract.

11.1.7 In *Monk Building and Civil Engineering Ltd* v. *Norwich Union Life Assurance Society* (1993) it was held that no contract had been concluded since *inter alia* several of the contract terms, including the liquidated damages provisions, had not been resolved. It was held that even if the liquidated damages provisions had been agreed Monk still considered certain other items had to be agreed before the contract could be finalised and both parties considered it essential that the final contract should be under seal. The fact that Monk had commenced work, relying on the contract provisions, was not relevant.

It was considered that there may be cases were a letter of intent provides a satisfactory basis for an 'if' contract and that it may sometimes be possible to imply terms which are missing from the letter of intent itself. However an 'if' contract must still contain the necessary terms. It was also held that it must be clear that the 'if' contract is to apply to the main contract work as opposed to preparatory work, if no formal agreement is ever reached.

11.1.8 A letter of intent can be in itself a mini contract. This was the situation in the case of *A.C. Controls Ltd* v. *BBC* (2002). A.C. Controls successfully tendered for the provision and installation of an integrated access control system at 57 of the BBC's premises across the UK. After completion of the tender process, ACC was informed that it had been successful but because the BBC was not in a position to issue a finalised contract, it sent ACC a letter of intent. This stated that the BBC intended to enter into a formal contract incorporating MF/1 conditions. The contract sum was £2 545 612 with an agreed addition of £115 000 to the preliminaries for extending the programme from 9 to 12 months. The letter of intent required ACC to carry out survey work and authorised expenditure up to a limit of £250 000.

The parties were in dispute concerning the letter of intent. ACC took the view that it was a contract in itself. On the other hand, the BBC argued that the letter of intent authorised but did not commit ACC to start work and that either party could terminate the arrangement at any time without notice. Payment due would be only for the work completed up to the point of termination. The court held that the letter of intent was an 'if' contract whereby if ACC accepted the offer contained in the letter of intent to carry out survey work, it would be obliged to carry out all the work provided for in the letter of intent and would be entitled to payment based upon a reasonable value of the work.

11.1.9 In some cases a letter of intent may represent only one of the documents which make up a contract. The case of *Twintec Ltd* v. *GSE Building & Civil Engineering Ltd* (2003) illustrates a complex situation involving a quotation, letter of intent, an acceptance letter and a meeting, which when read together formed the basis of the contract.

11.1.10 The meaning of the term 'subject to formal contract' was an issue in the Court of Appeal case of *Stent Foundations* v. *Carillion Construction (Contracts) Ltd* (2000). Normally, if negotiations are undertaken which are expressed as being 'subject to formal contract', the intention is that a contract will not come into being until a formal contract has been signed. In the *Stent* case it was held that a contract which was never formalised came into existence even though the letter of intent was 'subject to formal contract'. The court was of the opinion that, as the parties had agreed all the essential terms, a contract had been properly formed.

Similar situations arose in *Harvey Shopfitters* v. *ADI* (2004) and *Westminster Building Company* v. *Buckingham* (2004).

11.1.11 Other cases where the effects of letters of intent were an issue include:

Durabella Ltd v. *J. Jarvis & Sons Ltd* (2001)
Metropolitan Special Projects Ltd v. *Margold Services Ltd* (2001)
Shimizu Europe Ltd v. *L.B.J. Fabrications Ltd* (2003)
Tesco Stores Ltd v. *Costain Construction Ltd* (2003)
ABB Engineering & Construction Pty Ltd v. *Abigroup Contractors Ltd* (2003).

11.1.12 The moral of these cases is clear. Parties to a contract should make sure that all the terms are agreed before work commences. If this is not possible, any letter of intent should be adequately worded as to the precise method of payment in respect of any work requested to be carried out.

SUMMARY

The effect of a letter of intent is dependent upon the wording. If the contractor or subcontractor is instructed to commence design, order materials, or commence work and complies with the instruction he is entitled to receive fair and reasonable payment. If subsequently a contract is entered into the instruction will normally merge into the contract and payment will be made in accordance with the terms of the contract. If no contract is concluded payment will be on the terms of the letter of intent or on a quantum meruit basis.

11.2 What obligation does a contractor, subcontractor or supplier have to draw attention to onerous conditions in his conditions of sale?

11.2.1 There is extensive legislation protecting consumers from unfair and ill-considered contracts but these rules, and many of the provisions of the Unfair Contract Terms Act 1977, only apply to 'consumers', i.e. not to companies, or even persons, acting in the course of their business.

The general rule of *caveat emptor* will normally apply to a commercial contract – let the buyer beware. Courts constantly inform litigants that they do not have any power or obligation to correct contracts which are one-sided. Like all rules, however, there are exceptions. A court will from time to time intervene if it considers that a term in a contract is so onerous its presence should have been drawn to the attention of the other party to the contract at the time the contract was entered into. This will only apply in exceptional circumstances where one of the parties is using its own standard conditions.

11.2.2 It was held in the case of *Interfoto Picture Library Ltd* v. *Stiletto Visual Programmes* (1987) that the party seeking to enforce a particularly onerous set of printed conditions in a contract document had to demonstrate that the other party was sufficiently aware of the condition in question. If this were not shown, then the condition would not be incorporated into the contract.

The defendants failed to return some transparencies to the plaintiff agency and incurred a charge of £5 per day plus VAT per transparency. The sum in contention was £3783.50. It was held by the court that when the contract was entered into the plaintiff should have specifically drawn the defendant's attention to the charge for the late return of the negatives. Due to this failure the court reduced the charge from £5 per day per transparency to a more reasonable charge of £3.50 per transparency per week.

11.2.3 The case of *Worksop Tarmacadam Co Ltd* v. *Hannaby and Others* (1995) heard in the Court of Appeal dealt with an onerous term in a civil engineering contract. The plaintiffs were a small firm of civil engineering contractors, the defendants architects and engineers and other parties concerned in a housing development.

Roads and sewer works had to be undertaken for the development and the architects and engineers sought tenders for the same.

The contractor submitted a tender of £24 268.

- At the foot of the tender it stated 'N.B. THIS IS NOT A FIXED PRICE.'
- On the reverse were a number of conditions, clause 4(e) of which stated 'At any time before completion of the contract, the company shall be entitled to vary the price and take into account the following factors . . .'.
- Clause 15 provided that all work would be measured upon completion.

11.2.4   The plaintiffs entered into a contract and completed the works. During construction they encountered unforeseen ground conditions which necessitated extra work. On remeasurement the value of excavating the unforeseen ground was ascertained at a sum of £3007.

The court held that there was no entitlement to this amount and the appellant contractor appealed.

11.2.5   The Court of Appeal commented on the conditions of contract as follows:

> (1)   [The appellants'] submission is that clause 15, read literally and in context, is sufficiently wide to permit [them] to charge for the additional work that they encountered because of hard rock. I disagree. Had the plaintiffs wished to make such a provision in the event of unforeseen ground conditions being encountered, it would have been the easiest thing in the world for them so to have provided in specific terms. They did not do so.
>
> Thus, although perhaps by a somewhat different route, I reach the same conclusion as did Judge Bullimore; namely that the plaintiffs are not entitled to the additional £3007 because that cost is not caught by any of the conditions to be found on the reverse of the tender.
>
> (2)   From the learned County Court judge's point of view, this was not the end of the matter because, in a very extensive judgment, prepared with skill and care, the judge went on to make some specific findings in relation to whether clause 15 ever formed part of the contract between parties. He came to the conclusion that the plaintiffs were unable to rely upon it, not only on the grounds already indicated in this judgment, but because it had not been sufficiently brought to the defendants' attention. I would prefer not to express any definitive view on that aspect of the case but to proceed on the basis that the clause was brought to the attention, certainly of Mr Hannaby, who must be taken to have experience in this form of transaction.
>
> That said, however, although it is unnecessary for the reasons I have given in this judgment to make any express finding upon it, I take the view that clause 15, if it is not corollary to clause 4 and 8 providing machinery for ascertaining the true cost to the plaintiffs, was in any event so vague and in some ways so onerous a term that, without more specific attention being directed to its terms and those terms being brought specifically to the attention of Mr Hannaby, it did not form part of this contract.
>
> (3)   For all these reasons, and in agreement with the learned judge, although perhaps adopting a slightly different approach to the problems as he did, I have come to the conclusion that his ruling was correct that the plaintiffs were not, and are not, entitled to the additional £3007.

SUMMARY

The consumer protection legislation does not normally apply to commercial contracts. However where contractual terms are contained in one party's own standard terms and are considered by the court onerous they must be brought to the other party's attention at the time the contract is entered into otherwise they may be unenforceable. In the case of *Worksop Tarmacadam* v. *Hannaby* (1995) conditions on the reverse side of the quotation referred to all work being the subject of remeasurement. The conditions were not standard form conditions and on the face of the quotation it appeared to offer the work for a lump sum. The judge held that such a clause was unenforceable as it was not specifically drawn to the purchaser's attention.

11.3 Are there any restrictions on an architect/engineer's powers where the specification calls for the work to be carried out to the architect/engineer's satisfaction?

11.3.1 The standard forms in general use require the contractor to carry out the work to the reasonable satisfaction of the architect, engineer or supervising officer. For example, JCT 98 clause 2.1 refers to the contract documents indicating that 'the quality of materials or the standards of workmanship is a matter for the opinion of the Architect, and states 'such quality and standards shall be to the reasonable satisfaction of the Architect'.

11.3.2 Vincent Powell-Smith in *The Malaysian Standard Form of Building Contract* had this to say on the matter:

> 'There is no definition of what is meant by reasonable satisfaction anywhere in the contract. . . . "Reasonable satisfaction" might appear to suggest that the test is an objective one, but in truth the test is the subjective standards of the particular Architect, and there is a strong element of personal judgment in that opinion. It is reviewable in arbitration under clause 34 and the expression of satisfaction or otherwise by the Architect can be challenged by both the Employer and the contractor, provided a written request to concur in the appointment of an arbitrator is given by either party before the issue of the final certificate or by the contractor within 14 days of its issue: see clause 30(7).'

11.3.3 The manner in which an architect exercises his duties was examined in the case of *Sutcliffe* v. *Thackrah* (1974). In this case it was stated that the employer and the contractor enter into their contract on the understanding that in all matters where the architect has to apply his professional skill he will act in a fair and unbiased manner.

It would seem that under JCT 98 where the term 'reasonable satisfaction of the Architect' is used then the architect when exercising his powers, cannot demand standards which exceed those specifically referred to in the specification. Further, he must act in a fair and unbiased manner. Should the contractor or subcontractor be dissatisfied with the architect's decision then the remedy lies in a reference to an arbitrator who is given express powers under the arbitration clause to 'open up, review and revise any certificate, opinion, decision . . . [of the architect]'.

Disputes of this nature under JCT 98 may be referred under Article 5 to an adjudicator whose powers include, under clause 41A.5.2,

> 'opening up, reviewing and revising any certificate, opinion, decision, requirement or notice issued, given or made under the Contract . . .'

11.3.4 The ICE 6th and 7th Editions include slightly different words in clause 13(1) which states that

> 'the Contractor shall construct and complete the Works in strict accordance with the Contract to the satisfaction of the Engineer . . .'.

There is no reference to 'reasonable' satisfaction. Does this mean that the engineer has a greater discretion under the ICE conditions than that of an architect under JCT conditions? In all probability a court would hold that the contract contained an implied clause that the engineer in exercising his powers must act reasonably. A previous reference has been made to the decision in *Sutcliffe* v. *Thackrah* when it was held that an architect is obliged to act in a fair and unbiased manner. The same would apply to an engineer. It is submitted that in practical terms there is little difference between a contractor or subcontractor's obligation to carry out work to the 'satisfaction' or 'reasonable satisfaction' of the architect or engineer.

In like manner to an adjudicator or arbitrator appointed under a JCT form of contract the arbitrator appointed under the ICE conditions has power under clause 67(11) to

> 'open up review and revise any decision opinion instruction direction certificate or valuation of the Engineer.'

SUMMARY

There is no definition as to what is meant by reasonable satisfaction. It might appear to suggest that the test is an objective one, but in truth the test is the subjective standards of the particular architect or engineer. If the contractor is not satisfied his recourse is to refer the matter to adjudication or arbitration.

11.4 If an estimate prepared by an engineer or quantity surveyor proves to be incorrect can the employer claim recompense?

11.4.1 For an employer to recover from an engineer or quantity surveyor as a result of an incorrect estimate it will be necessary to prove breach of an obligation and a resultant financial loss.

The consultancy agreement between the employer and engineer or quantity surveyor can take many forms. It may be that a specially drafted agreement is used or a standard ICE or RICS agreement is employed. A simple exchange of letters or even a verbal agreement may be the basis of the contract. The terms of the contract may be express or implied. It is normal for a contract such as this to indicate that the engineer or quantity surveyor will have an obligation to carry out his duties with reasonable skill and care. If there is no express term to this effect then one will normally be implied by law.

In the case of *Bolam* v. *Friern Hospital Management Committee* (1957) the House of Lords had to define reasonable skill and care. Mr Justice McNair had held that it was not necessary to achieve the highest possible professional standards:

> A man need not possess the highest expert skill at the risk of being found negligent. It is well established law that it is sufficient if he exercised the ordinary skill of an ordinary competent man exercising that particular art.

11.4.2 The employer may as an alternative when seeking to recover costs resulting from an incorrect estimate seek to demonstrate that the error constituted a breach of warranty. In the case of *Copthorne Hotel (Newcastle) Ltd* v. *Arup Associates* (1997) an action was brought by an employer against one of its consultants as a result of an incorrect estimate on such a basis.

In 1987 the plaintiff decided to construct a hotel in Newcastle and after an initial abortive design with a local architect the plaintiff engaged Arup as engineers, architects and quantity surveyors for the project.

Arup produced a number of budgets or estimates. Rush & Tompkins were engaged as contractor but went into receivership and the hotel was completed by Bovis Construction. Practical completion was achieved in February 1991 for a total sum of £15 205 000.

The plaintiff withheld fees from Arup as they were dissatisfied with their work and consequently Arup terminated the contract in July 1991. In 1994, the plaintiff commenced court action claiming that Arup had given inaccurate cost estimates and had failed to design within the construction costs estimated. Further, they had failed to control costs, co-ordinate the work or supply prompt information, and there were defects in the works. Arup counterclaimed for fees allegedly owed.

In 1995 the Official Referee directed that the trial should be of specified questions in relation to certain paragraphs in the statement of claim and the defence and counterclaim. These included, under liability proper:

- the existence and scope of duty
- whether there were any breaches, and if so, what breaches.

Paragraph 11 of the statement of claim read:

'At a meeting held in mid April 1988, attended by Norman Cooke and Kenneth Hunt of the plaintiff and Peter Foggo, Michael Lowe and Dick Lee of the defendant, the defendant confirmed that the hotel to be designed by the defendant could be completed for the sums proposed by the plaintiff with a margin of +/- 5%, namely, an overall cost of £12 m, of which sum the construction costs would be £8 m approximately.'

The overall cost of the hotel proved to be £21.2 m, of which building costs accounted for £15 205 000.

11.4.3 In finding in favour of Arup Judge Hicks QC said:

Paragraph 11 uses the word 'confirmed', but there is no suggestion of any prior statement susceptible of confirmation so it should presumably be understood as meaning 'stated'. However that may be, there is certainly no use of any technical term such as 'represented', 'advised' or 'warranted' to indicate what is alleged as to the legal status of the words used or as to the nature of the relationship between the parties. In paragraphs 22, 23 and 24, however, what is alleged in paragraph 11 is referred to as 'advice'. Although I have no doubt, as I have indicated ... above, that Arup was offered and accepted within a few days of 12 April 1988 the appointment which became much later the subject of a contract in writing, I am equally clear that there was no contractual relationship between the parties at the meeting referred to in paragraph 11 of the re-amended statement of claim.

> Whatever was said then was not, therefore, professional advice given pursuant to a contract. The most natural analysis, in my view, is that, if anything of the kind alleged was said, it was a representation intended to induce the plaintiff to engage Arup. That is not expressly pleaded.

In other words, there may possibly have been a misrepresentation inducing the employer to enter into the contract, but there was not any warranty which could be said to have been breached by the cost overrun as there was at that time no contract.

11.4.4 In the Singapore case of *Paul Tsakok & Associates* v. *Engineer & Marine Services (Pte) Ltd* (1991) an architect brought an action against a client for unpaid fees.

The architect had given an estimate for a project to the client. Eventually the client decided not to call any tenders because he considered that tenders would exceed the estimate but instead ordered the architect to revise all the drawings so that the tenders would come within the estimate given. The architect declined and demanded his fees. The client refused to pay.

The court held that it was never the parties' intention that the estimate was guaranteed. The architect could not be held responsible in this case and was entitled to be paid the proper full fees pursuant to the SIA terms of engagement.

This case shows that consultants are not expected to be perfect and that clients cannot dismiss them as and when they so desire. The architect, quantity surveyor or engineer does not normally guarantee an estimate he gives to a client. However, although the court found in favour of the architect, the judge considered that both parties had behaved unreasonably and the matter could have been resolved without going to court. Because of this he only awarded the architect one half of his legal costs.

SUMMARY

An estimate which proves to be incorrect in itself will not provide the employer with a right of redress. To be successful the employer must demonstrate that the engineer or quantity surveyor warranted the accuracy of the estimate or show that it was exceeded due to a lack of reasonable skill and care on the professional's part. It is possible for the estimate to be inaccurate for reasons outside the control of the engineer or quantity surveyor, for example, a change in market conditions.

Further, if it can be shown that had the estimate been correct the employer would in any event have proceeded with the project, the recoverable loss will be nominal even though an action is otherwise successful. This is because the employer, in continuing with the scheme, will be unable to demonstrate loss as the incorrect estimate did not affect his decision. If, on the other hand, once tenders are received above the estimate the scheme is abandoned, the employer should be able to demonstrate loss.

11.5 When defects come to light after the architect/engineer has issued the final certificate, does the contractor/subcontractor still have a liability or can he argue that once the certificate has been issued the employer loses his rights?

11.5.1 The Limitation Act 1980 provides periods of time within which actions must be commenced after which they are 'statute barred', providing the defendant with an unassailable defence.

Section 7 states:

> 'An action to enforce an award, where the submission is not by an instrument under seal, shall not be brought after the expiration of six years from the date on which the cause of action accrued.'

Section 8 states:

> An action upon a speciality [i.e. a contract under seal or expressed as a deed] shall not be brought after the expiration of twelve years from the date on which the cause of action accrued.

With construction contracts where a breach has occurred, for example defective work, those periods will usually begin to run from the date of practical completion. Claimants therefore have six years or, in the case contracts under seal or expressed as a deed, twelve years after practical completion within which to commence a court action.

11.5.2 Questions are regularly asked as to how these periods are affected by the issue of a final certificate. Does the issue of a final certificate shorten the periods of limitation set down by the Limitation Act?

11.5.3 With regard to JCT contracts, JCT 80 prior to the issue of amendment 15 in July 1995 makes reference in clause 30.9.1 to the final certificate being conclusive evidence as to the architect's satisfaction where it states:

> '... the Final Certificate shall have effect in any proceedings arising out of or in connection with this Contract (whether by arbitration under Article 5 or otherwise) as ... conclusive evidence that where and to the extent that the quality of materials or the standard of workmanship is to be of the reasonable satisfaction of the Architect the same is to such satisfaction.'

An exception would occur if arbitration proceedings were commenced either before the issue of the final certificate or within a period up to 28 days after its issue.

The Court of Appeal gave a decision on the effect of that clause in the case of *Crown Estate Commissioners* v. *John Mowlem and Co Ltd* (1994). It was held that the final certificate would safeguard a contractor from any claims made after its issue.

Sir John Megaw had this to say in giving judgment:

> Where the final certificate thus becomes conclusive evidence, the effect is that any claim in an arbitration which seeks to support some provision of the final certificate is bound to succeed, and any claim which seeks to challenge a provision of the final certificate is bound to fail, without any hearing on the merits.

11.5.4   The effect of the *Crown Estate* case was that once the final certificate had been issued and a further 28 days had elapsed either party could commence an action but could not produce evidence to support its case. Employers in particular were placed in a less advantaged position in using JCT contracts than would have been the case if other forms of contract had been employed. The drafters of the JCT forms therefore issued amendment 15 in July 1995 to negate the effect of the *Crown Estate* case:

> '... but such Certificate shall not be conclusive evidence that such or any other materials or goods or workmanship comply or complies with any other requirement or term of this Contract...'

The final certificate remains conclusive in so far as it relates to qualities of materials or goods or any standard workmanship expressly described in the contract documents to be to the approval of the architect.

The Scottish courts in *Belcher Food Products Ltd* v. *Miller & Black and Others* (1998) gave a different view to the meaning of clause 30.9.1 of JCT 80 for amendment 15. It was the opinion of the Scottish Outer House in respect of a defective floor screed that the decision in *Crown Estates* could prove unfair to the employer. It was the court's decision that the final certificate is conclusive only in respect of those matters expressly reserved by the contract to be to the architect's satisfaction.

11.5.5   In the case of *London Borough of Barking & Dagenham* v. *Terrapin Construction Ltd* (2000), the Court of Appeal was asked to address the matter in respect of a contract let under a JCT 81 With Contractor's Design form. Clause 30.8.1.1 of the With Contractor's Design is similar to clause 30.8.1 of JCT 80 in that it states 'as conclusive evidence that where it is stated in the Employer's Requirements that the quality of material or the standard of workmanship are to be to the reasonable satisfaction of the Employer the same are to such satisfaction'. It was argued on behalf of the employer that in the particular case, as there was no specific document referred to as embodying the Employer's Requirements, the clause was inoperable. Further, unlike the JCT 80 contract where there was an architect who had to be satisfied, there was no like person under the With Contractor's Design form.

The Court of Appeal was unimpressed. It was held that clause 1 of the agreement required the execution, completion and maintenance of the works to be in all respects to the satisfaction of the employer. Further, the employer may choose to perform the functions assigned to him personally or through a qualified agent. The conclusiveness applied to all defects whether latent or patent. The wording of clause 30.8.1.1 refers to defective goods or materials but does not lock out claims resulting from defective design. It was also held that the wording of clause 30.8.1.1 does not exclude claims based upon any failure to meet any statutory requirements which, under the provisions of the contract, are the responsibility of the contractor.

11.5.6   The Model Form MF/1 also refers to the final certificate as being conclusive – with exceptions.

Clause 39 states:

'A final certificate of payment shall be conclusive evidence:

- that the works or section to which such certificate relates is in accordance with the contract;
- that the contractor has performed all his obligations under the contract in respect thereof; and
- of the value of the works or section.

Payment of the amount certified in a final certificate of payment shall be conclusive evidence that the purchaser has performed all his obligations under the contract in relation to the works or section thereof to which the certificate relates.

A final certificate of payment shall not be conclusive as to any matter dealt with in the certificate in the case of fraud or dishonesty relating to or affecting any such matter.

A final certificate of payment shall not be conclusive if any proceedings arising out of the contract whether under clause 52 [Disputes and Arbitration] or otherwise shall have been commenced by either party in relation to the works or section to which the certificate relates,

- before the final certificate of payment has been issued, or
- within three months thereafter.

11.5.7 The Institution of Chemical Engineers' Model Conditions of Process Plant 1981 Edition, the Red Book, at clause 38.5 includes the wording:

'The issue of the final certificate for the plant as a whole or, where for any reason more than one final certificate is issued in accordance with this clause, the issue of the last final certificate in respect of the works, shall constitute conclusive evidence for all purposes and in any proceedings whatsoever between the purchaser and the contractor that the contractor has completed the works and made good all defects therein in all respects in accordance with his obligations under the contract...'

The effect of this was subject to dispute in *Matthew Hall Ortech Ltd* v. *Tarmac Roadstone Ltd* (1997) in respect of a contract for the design, erection and commissioning of a mixed processing plant.

Matthew Hall carried out the work and it was alleged by Tarmac Roadstone, the other party to the contract, that 22 steel bunkers were suffering from what appeared to be structural damage as a result of design and construction deficiencies emanating from Matthew Hall's breaches of contract. The responsibility for issuing the final certificate lay with Tarmac Roadstone, but no such certificate was ever issued.

Matthew Hall argued that a final certificate should have been issued, and, if it had been, it would have acted as a bar to any claim from Tarmac Roadstone. It was Tarmac's argument that, even if they had issued a final certificate, they would not be prevented from bringing a claim against Matthew Hall.

The matter was referred to arbitration. It was held by the arbitrator that a final certificate would bar contractual claims against Matthew Hall for defects which it was accused of not correcting, but would not prevent either a contractual claim for defects allegedly put right, but later discovered to have been done badly, or a claim for latent defects.

The court did not fully agree with the arbitrator. It was considered that there would appear to be commercial justification for the contract to provide a definite cut-off point once plant had been constructed, tested, provided and made good in all respects in conformity with the contract. The court considered that Matthew Hall were correct in their contention that a final certificate is conclusive evidence that all work has been completed in accordance with the requirements of the contract. It was also held that the absence of a final certificate when one ought to have been issued would not be allowed to defeat the object of clause 38.5.

11.5.8   The ICE 6th and 7th Editions and GC/Works/1 1998 conditions do not make the final certificate in any way conclusive therefore the periods laid down in the Limitation Act within which actions must be commenced apply.

The ICE Conditions at clause 61(2) put it very clearly

> 'The issue of the Defects Correction Certificate shall not be taken as relieving either the Contractor or the Employer from any liability the one towards the other arising out of or in any way connected with the performance of their respective obligations under the Contract.'

Clause 39 of the GC/Works/1 1998 is not so emphatic, but it makes no reference to the certificate being final.

11.5.9   In the decisions in *Gray* v. *TP Bennett & Son* (1987) and *William Hill Organisation Ltd* v. *Bernard Sunley & Sons Ltd* (1982) it was held that a final certificate could not be conclusive where, due to fraudulent concealment, the defects could not have been detected following reasonable inspection.

SUMMARY

JCT 80 (in England but not Scotland) prior to amendment No. 15 issued in July 1995, MF/1 and the IChemE Forms make a final certificate conclusive evidence that work has been satisfactorily carried out. Actions under the JCT wording must be commenced within 28 days of the issue of the final certificate and in the case of MF/1 within three months to be effective.

Other commonly used standard forms of contract give the final certificate no such effect and the Limitation Act periods apply.

JCT amendment No. 15 is now incorporated in JCT 98 and the final certificate is conclusive only in respect of particular qualities of materials or goods, or any standard of workmanship expressly described in the contract documents as being to the approval of the architect.

A final certificate is not conclusive if due to fraudulent concealment the defect could not have been detected following reasonable inspection.

11.6 Where a nominated subcontractor becomes insolvent after the nominated subcontract works have been completed, will the architect/engineer be required to nominate another subcontractor to remedy the defects with the employer bearing the cost, or will the main contractor 'pick up the bill'?

11.6.1 This matter is all a question of who bears the risk for failure by a nominated subcontractor – the employer or main contractor. Most standard forms of contract make it clear.

11.6.2 GC/ Works/1 1998 in condition 63A lays the responsibility firmly on the shoulders of the employer.

The clause covers the situation where the subcontract is determined or assigned due to the subcontractor's insolvency. Under this clause the contractor can recover from the employer any additional cost incurred in completing the subcontract work in excess of what he might have recovered using his best endeavours from the subcontractor.

Main contractors should ensure that under the terms of the subcontract the subcontract is automatically determined where the subcontractor becomes insolvent. Additional cost claims should be made against the insolvent subcontractor even if there is little chance of receiving any payment.

11.6.3 Clause 7.2 of nominated subcontract NSC/C for use with the JCT 98 makes it clear that if the nominated subcontractor becomes insolvent its employment is automatically determined. It will be for the main contractor to notify the architect in accordance with clause 35.24.2 that this has happened. Once the subcontractor's employment has been determined the architect will nominate a replacement subcontractor as required by clause 35.24.6.3.

The purpose of these provisions is fully to compensate the main contractor for any costs resulting from default on the part of a nominated subcontractor.

11.6.4 The ICE 6th and 7th Editions in similar fashion to GC/Works/1 1998 and JCT 98, place the risk of default by a nominated subcontractor upon the employer.

Clause 59 (4)(a) refers to the nominated subcontract being terminated. The engineer's consent is required by clause 59 (4)(b) to expel the subcontractor from the site. Renomination is covered by clause 59 (4)(c) referring back to clause 59(2).

There is an obligation placed upon the main contractor by clause 59 (4)(d) to take all reasonable steps to recover any additional cost from the nominated subcontractor. Where there is a shortfall the employer is required by clause 59 (4)(e) to reimburse the contractor.

SUMMARY

The standard forms of contract GC/Works/1 1998, JCT 98 and ICE 6th and 7th Editions all place the risk of defective work which is not corrected by a nominated subcontractor who has become insolvent on the employer.

GC/Works/1 1998 and ICE 6th and 7th Editions place the onus on the main contractor to recover the additional costs from the defaulting subcontractor but any shortfall is to be made good by the employer.

11.7 Can the contractor in his order form impose conditions that take precedence over any conflicting conditions referred to in the subcontractor's acceptance?

11.7.1 There are three necessary elements to a binding contract:

- Agreement
- Contractual intention
- Consideration.

11.7.2 Agreement often manifests itself when an unambiguous offer receives an unconditional acceptance. It is not uncommon for an offer to be met with a counter-offer which may or may not be accepted. In an effort to gain an advantage it is not unknown for a counter-offer to be met with a counter-counter-offer. This is often referred to as 'the battle of the forms'. The classic case involving this principle is *Butler Machine Tool Co Ltd* v. *Ex Cell-O Corporation* (1979).

In this case sellers offered to supply a machine for a specified sum. The offer was expressed to be subject to certain terms and conditions, including a 'price escalation clause'. In reply the buyers placed an order for the machinery on their own terms and conditions, which differed from those of the sellers in containing no price escalation clause and also in various other respects. It also contained a tear-off slip to be signed by the sellers and returned to the buyers stating that the sellers accepted the order 'on the terms and conditions stated therein'. The sellers did sign the slip and returned it with a letter saying that they were 'entering' the order 'in accordance with' the offer. This communication from the sellers was held to be an acceptance of the buyers' counter-offer so that the resulting contract was on the buyers' terms, and the sellers were not entitled to the benefit of the price escalation clause.

This is sometimes referred to as the 'last shot' principle – in other words the party who fires the last shot wins the battle.

11.7.3 The question above deals with a contractor in placing an order with a subcontractor trying to head off any counter-offer that might be introduced in the subcontractor's own terms. This is done by stating in the order

> 'unless otherwise agreed the subcontractor is deemed to have accepted those conditions ... which shall apply to the exclusion of any conditions which appear on the acceptance form.'

It is clear that the subcontractor's terms will not apply unless agreed by the main contractor.

In like manner to the *Butler Machine Tool* case the subcontractor may send an acceptance with terms included which vary from those in the main contractor's order and have a tear-off slip to be signed by the main contractor stating that the contractor agrees to the terms set out in the acceptance. If the main contractor, on receiving the acceptance, signs and returns the tear-off slip he will have 'otherwise agreed' to the subcontractor's terms.

11.7.4   A more difficult situation would arise if, having received the contractor's order, the subcontractor forwarded his written acceptance which makes reference to the subcontractor's own terms but with no tear-off slip.

The contractor may on receiving the acceptance instruct the subcontractor to commence work having made no further reference to the order or the acceptance. However, an offer or counter-offer may be accepted by conduct. For example, in *Brogden v. Metropolitan Rly Co* (1877) a railway company submitted to a merchant a draft agreement for the supply of coal. He returned it marked 'approved' but also made a number of alterations to it, to which the railway company did not expressly assent. Nevertheless the company accepted deliveries of coal under the draft agreement for two years. It was held that once the company began to accept these deliveries there was a contract on the terms of the draft agreement as amended.

If the contractor in the question, having received the subcontractor's acceptance without comment, gives an instruction to commence, has he accepted by conduct the conditions stated in the acceptance?

11.7.5   The court in deciding the issue may take into account the actions of the parties subsequent to the commencement of work. Were they operating on the terms included in the order or the acceptance? If this is not conclusive the court may take the view that the contractor made it clear at the outset that he would not accept any terms included in the subcontractor's acceptance unless otherwise agreed. The subcontractor would have to show very clearly that, despite the wording of the contractor's order, an agreement in the terms of the acceptance had been reached.

11.7.6   In the case of *Goodmarriot and Hursthouse* v. *Young Austin & Young Ltd* (1993) the judge was influenced by wording of this nature and considered it relevant that the defendant had stipulated in the order the means by which a concluded contract was to come into existence.

SUMMARY

A contractor is entitled in his order specifically to state that he is not prepared to accept terms in a subcontractor's acceptance or confirmation of order unless he otherwise agrees.

Where wording such as that set out in 11.7.3 is used, if the terms in the subcontractor's acceptance or confirmation of order vary from those in the order the subcontractor must show that they have been specifically accepted by the contractor if they are to apply.

11.8 Who is responsible if damage is caused to a subcontractor's work by person or persons unknown – the subcontractor, contractor or employer?

11.8.1 Contractors normally like to pass down to subcontractors the risk of damage to the subcontract works. Non-standard subcontracts are often worded in such a manner that the subcontractor is expressly required to protect the subcontract works to prevent damage. Courts will be obliged, where a dispute arises, to place an interpretation on such wording.

In the case of *WS Harvey (Decorators) Ltd* v. *HL Smith Construction Ltd* (1997) the terms of the subcontract required the subcontractor to provide 'all necessary and proper protection'.

The court held that all necessary protection means such protection as is necessary to prevent damage to the works from whatever cause. Further the clause stated that the subcontractor:

> 'will be held responsible for the adequacy of the protection afforded and shall make good or re-execute any damaged work at his own expense.'

The judge in finding for the contractor said that the wording imposed the obligation of protecting the works firmly and squarely upon the subcontractor.

11.8.2 Damage is categorised under three headings where the standard subcontract forms for use with JCT 98 apply:

- That caused by the specified perils, e.g. fire, storm, tempest etc. is an insurance risk for which the contractor or employer will be liable by clause 8C.2.1 of DOM/1.
- That caused by any negligence, omission or default of the contractor, his servants or agents or any other subcontractor will be the responsibility of the main contractor under clause 8C.2.1 of DOM/1.
- Where materials or goods have been fully, finally and properly incorporated into the works but before practical completion of the subcontract works the contractor will be responsible under clause 8C.2.2 of DOM/1.

Some difficulties have been experienced in deciding when the main contractor becomes liable under the third category above. In particular an interpretation of the wording 'fully, finally and properly incorporated into the works' is required.

11.8.3 It has been argued by some main contractors that this stage cannot be achieved until all the subcontract works have been completed and accepted on behalf of the employer. This cannot be correct as clause 8.3.2 refers to materials or goods having been fully, finally and properly incorporated into the works *before* practical completion of the subcontract works.

The wording obviously contemplates the stage being reached before practical completion of the subcontract and hence this argument does not hold good. The essence is therefore the wording 'fully, finally and properly incorporated into the works'.

The works are defined in DOM/1 as 'the main contract works including the subcontract works'.

In the *Concise Oxford Dictionary* the remainder of the words are defined as:

fully   –   completely – without deficiency
finally   –   coming last
properly   –   suitably, rightly

11.8.4 A reasonable interpretation would therefore be that liability lies with the main contractor for damage to the subcontractor's materials or goods once they have been put without deficiency (i.e. all in place and nothing missing) in their final position and are suitable in respect of the contract requirements.

With regard to materials manufactured off site, for example ceiling tiles or wall tiles, when they are fixed in position with nothing further to be done to them and comply with the requirements of the contract then they are 'fully, finally and properly' incorporated into the works.

Where wet trades are involved such as plaster, paint or asphalt, once the wet material has been applied or laid and dried off then the materials or goods, if defect free, are fully, finally and properly incorporated into the works.

The purpose behind the wording would seem to be that as the subcontract works progress and parts of the work are completed, the subcontractor will move on leaving the completed parts behind. These completed parts become the responsibility of the main contractor as he and his following trades will by then be working in those completed areas.

Other subcontract forms for use with a JCT form such as NSC/C, NAM/SC, IN/SC and the like are worded in similar fashion to DOM/1.

11.8.5 The CECA Blue Form for use with the ICE 6th and 7th Editions is worded along different lines. Clause 14 provides for insurance to be taken out in accordance with the requirements of the Fifth Schedule for the risks set out therein. The wording of clause 14(2) states:

> 'The contractor shall maintain in force until such time as the main works have been substantially completed or ceased to be at his risk under the main contract, the policy of insurance specified in Part II of the Fifth Schedule hereto. In the event of the subcontract, or any subcontractor's equipment, temporary works, materials or other things belonging to the subcontractor being destroyed or damaged during such period in such circumstances that a claim is established in respect thereof under the said policy, then the subcontractor shall be paid the amount of such claim, or the amount of his loss, whichever is the less, and shall apply such sum in replacing or repairing that which was destroyed or damaged.'

It will be necessary for the subcontractor to make sure that Part II of the Fifth Schedule fully covers damage from all causes to the subcontractor's materials and equipment.

In the absence of adequate wording in the Fifth Schedule clause 14(2) places the risk on the subcontractor's shoulders in the following terms:

> 'Save as aforesaid the subcontract works shall be at the risk of the subcontractor until the main works have been substantially completed under the main contract, or if the main works are to be completed by sections, until the last of the sections in which the subcontract works are comprised has been substantially completed, and

the subcontractor shall make good all loss of or damage occurring to the sub-contract works prior thereto at his own expense.'

SUMMARY

Main contractors where non-standard forms are used like to include terms which place the risk of damage to the subcontract works onto the sub-contractor.

Where the standard forms of subcontract conditions for use with the JCT main conditions apply the subcontractor is liable for damage to goods until such time as they are 'fully, finally and properly incorporated into the Works' unless the damage has been caused by the specified perils (fire, storm, tempest, etc.) or due to negligence by the main contractor or other subcontractors.

In the case of the CECA Blue Form of subcontract the subcontractor is at risk until the main contract works have been substantially completed unless Part II of the Fifth Schedule states the contrary.

11.9   How is the term 'regularly and diligently' as used in the standard forms of contract to be defined?

11.9.1   The employer under most standard forms of contract is entitled to determine the contractor's employment if he fails to proceed regularly and diligently with the works. A definition of the term 'regularly and diligently' has been provided in the case of *West Faulkner Associates* v. *The London Borough of Newham* (1992).

11.9.2   An action was commenced by the architects for the recovery of fees and damages for wrongful repudiation of their contract. By way of defence and counterclaim the council alleged among, other things, default by the architects in not serving a default notice to Moss, the main contractor, under clause 25(1)(b) concerning a failure to proceed 'regularly and diligently' with the works.

If such a notice had been served it would have given rise to the council's entitlement to determine. The failure by the architects to serve the notice left the local authority powerless to effect a determination. Instead they were obliged to pay a substantial sum to the contractors for them to leave the site. The court had to decide the meaning of the words 'regularly and diligently' and whether in fact the contractor had failed to proceed in that manner. Judge John Newey QC, having listened to expert evidence and had his attention called to various precedents, decided:

'In the light of the judgments, textbooks and expert evidence I conclude that regularly and diligently should be construed together and that in essence they mean simply that contractors must go about their work in such a way as to achieve their contractual obligations. This requires them to plan their work, to lead and to manage their workforce, to provide sufficient and proper materials and to employ competent tradesmen, so that the works are fully carried out to an acceptable standard and that at all time, sequence and other provisions of the contract are fulfilled.'

11.9.3 Judge Newey concluded that Moss did not plan their work properly, did not provide efficient leadership or management and some at least of their tradespeople were not reasonably competent and therefore they had failed to proceed regularly and diligently with work. It was Judge Newey's opinion that Moss' failures were very extreme and the architects should have realised that Moss were not proceeding regularly and diligently and therefore served the notice.

As a direct result of the architects' failure to serve Moss with the notice, the council were prevented from terminating the contractor's employment. The council suffered loss by having to pay the new contractors more than they would have had to pay Moss. They also incurred additional costs in respect of site supervision, additional quantity surveyors' fees, payments to tenants and lost rent. All these losses Judge Newey considered flowed from the breach by the architects.

SUMMARY

Contractors who are required to carry out work regularly and diligently must go about their work in such a way as to achieve their contractual obligations. This requires them to plan their work, to lead and manage their workforce, to provide sufficient and proper materials and to employ competent tradesmen so that the works are fully carried out to an acceptable standard and that at all contractual time, sequence and other provisions are fulfilled.

11.10 Are there any circumstances under which a contractor/ subcontractor could bring an action for the recovery of damages against an architect/engineer for negligence?

11.10.1 A failure on the part of an architect/engineer to exercise reasonable skill when issuing payment certificates or performing other functions under the contract could prove expensive for the contractor. Does this leave the contractor with a right to recover his losses from the negligent architect/ engineer?

11.10.2 The matter was considered in the case of *Arenson* v. *Arenson* (1977) when Lord Salmon said:

> The Architect owed a duty to his client, the building owner, arising out of the contract between them to use reasonable care in issuing his certificates. He also, however, owed a similar duty of care to the contractor arising out of their proximity: see *Hedley Byrne & Co Ltd* v. *Heller & Partners Ltd* (1964), *Sutcliffe* v. *Thackrah* (1974).

In *Michael Salliss & Co Ltd* v. *Calil* (1987) Judge Fox-Andrews also held that an architect held a duty to the contractor to act fairly when certifying:

It is self-evident that a contractor who is party to a JCT contract looks to the architect or supervising officer to act fairly as between him and the building employer in matters such as certificates and extensions of time. Without a confident belief that that reliance will be justified, in an industry where cash flow is so important to the contractor, contracting would be a hazardous operation. If the architect unfairly promotes the building employer's interest by low certification or merely fails properly to exercise reasonable care and skill in his certification it is reasonable that the contractor should not only have the right as against the owner to have the certificate reviewed in arbitration but also should have the right to recover damages against the unfair architect.

11.10.3 In *Pacific Associates* v. *Baxter and Halcrow* (1988) the Court of Appeal took a different view. Pacific entered into a contract with the Ruler of Dubai for the dredging of a lagoon in the Persian Gulf. Halcrow was appointed as the engineer. The contract incorporated the FIDIC conditions (2nd Edition, 1969). Condition 86 of the contract provided as follows:

> 'Neither any member of the Employer's staff nor the Engineer nor any of his staff, nor the Engineer's representative shall be in any way personally liable for the acts or obligations under the contract, or answerable for any default or omission on the part of the Employer in the observance or performance of any of the acts matters or things which are herein contained.'

The work was delayed because of the presence of hard materials. Pacific made claims for extensions of time and additional expenses which were rejected by Halcrow. Pacific then made a formal submission in accordance with condition 67 for the decision of Halcrow. When this too was rejected Pacific referred its claims to the ICC for arbitration in accordance with condition 67. The proceedings were compromised when the Ruler of Dubai agreed to pay Pacific some £10m in full and final settlement of its claims against him.

In March 1986 Pacific issued a writ claiming £45 m from Halcrow, being the unrecovered balance (including interest) of its claim against the Ruler of Dubai.

The Court of Appeal held:

(1) In considering whether a duty of care existed it was relevant to look at all the circumstances, and these included the contract between the Ruler of Dubai and Halcrow.
(2) There had been no 'voluntary assumption of responsibility' by Halcrow relied upon by Pacific sufficient to give rise to a liability to Pacific for economic loss in circumstances in which there was an arbitration clause. The position might well have been otherwise if the arbitration clause or some provision for arbitration had not been included in the contract.

SUMMARY

It would seem from the decision in *Michael Salliss & Co Ltd* v. *Calil* (1987) that architects and engineers could have a liability to a contractor if they fail to act fairly.

However the Court of Appeal decision in *Pacific Associates* v. *Baxter* and *Halcrow* (1988) went against the contractor as it was held that there was no evidence that the engineer had assumed any responsibility towards the contractor. Leaving architects and engineers open to claims from contractors could prove to be an unsatisfactory manner of carrying out construction works. Courts may in future cases continue to play the lack of 'voluntary assumption of liability' card to resist contractors' claims.

11.11 What is a contractor's liability to the employer for failing to follow the specification where it is impractical to take down the offending work?

11.11.1 Most of the standard forms of contract provide the architect or engineer with power to instruct the contractor to take down and remove work which does not comply with the contract. These powers extend beyond practical completion into the defects period.

A situation may arise where work is defective but it is not practical for it to be corrected. JCT 98 clause 17.3 provides for an appropriate deduction to be made in respect of defects, shrinkages or other faults which are not required to be made good.

11.11.2 The question of what level of recompense an employer would be able to recover from a contractor whose work was defective arose in the House of Lords case of *Ruxley Electronics and Construction Ltd* v. *Forsyth Laddingford Enclosures* (1995).

The dispute concerned the construction of a swimming pool, the maximum depth to which the pool was constructed being 6 feet 9 inches which differed from the 7 feet 3 inches depth specified. The respondent sought damages for breach of contract for the cost of demolishing the existing pool and rebuilding it to the required depth.

The trial judge had found that the pool as constructed was safe to dive into and that the deficiency had not decreased the value of the pool. Further, he was not satisfied that the respondent intended to rebuild the pool and that the cost of rebuilding would be wholly disproportionate to the disadvantage of having a pool which was less than a foot too shallow. Only damages for loss for amenity of £2500 were awarded.

The Court of Appeal allowed an appeal, finding that the only way the employer could have achieved the object of the contract was to reconstruct the pool at a cost of £21 560 and that this was reasonable. The contractors appealed.

It was held by the House of Lords that the award of damages was designed to compensate for an established loss and not to provide a gratuitous benefit to the aggrieved party. Therefore it followed that the reasonableness of the award was linked directly to the loss sustained. It was unreasonable to award damages for the cost of reinstatement if the loss sustained did not extend to the need to reinstate. A failure to achieve a contractual objective does not necessarily mean that there is a total failure. In the instant case, the employer had a perfectly serviceable pool, even if it was not as deep as it should have been. His loss was not the lack of a useable pool and there was no need to construct a new one. Reinstatement was not the correct measure of damages in this case. The claimant was therefore entitled to £2500 for loss of amenity only.

Lord Jauncey offered an *obiter* comment to the effect that in the normal case, the court has no concern with the use to which a plaintiff puts an award of damages for a loss which has been established. Intention, or lack of it, to reinstate can have relevance only to reasonableness and, hence, to the extent of the loss which has been sustained. Once that loss has been established, intention as to the subsequent use of the damages ceases to be relevant.

11.11.3 A case which deals with a similar matter is *RJ Young* v. *Thames Properties* (1999). The work provided for the construction of a car park. 100 mm of limestone should have been laid but instead the contractor provided only 30 mm. It was held by the court that if the contractor undertakes work which departs from what is required by the contract, there is nevertheless an entitlement to payment for what work was actually carried out unless the work was of no benefit, entirely different or incomplete.

The measure of damage was based upon the value of the car park as laid and not the cost of correcting the work.

11.11.4 The Scottish case of *McLaren Murdock and Hamilton* v. *The Abercromby Motor Group* (2002) had to deal with a defective heating system and the amount of money which should be paid by the offending party.

The judge applied the principle that the innocent party is generally entitled to be placed in as good a position financially as it would have been had the breach not occurred. The innocent party is entitled to recover the cost of remedial work unless these costs are disproportionate to the benefit to be gained from the remedial works. It was held that the cost of replacing the heating system was not disproportionate to the benefit to be obtained.

SUMMARY

The House of Lords decision gives authority to the view that the employer will be entitled to recover the cost of rectification if work carried out by a contractor is defective. However, if rectification was not a reasonable solution due to the high cost compared with minimal benefit, rectification costs would not be awarded. An award based upon loss of amenity may be more appropriate.

11.12 Do retention of title clauses still protect a supplier or subcontractor where a main contractor becomes insolvent or have there been cases which throw doubt on their effectiveness?

11.12.1 Where a company or organisation becomes insolvent there is usually insufficient money available for all to whom payment is due.

In an effort to protect themselves, some suppliers and subcontractors include in their terms of trading what has become known as a 'retention of title' clause. In essence the clause states that the goods supplied to the purchaser remain in the ownership of the supplier until payment has been made in full. In the event of a failure to make full payment the clause usually provides for their return. If, therefore, a purchaser becomes insolvent before paying for the goods, the liquidator or receiver should either make payment or allow the goods to be returned to the seller.

11.12.2 The basic principle is contained in the maxim *nemo dat quod non habet* and this is embodied in the Sale of Goods Act 1979, section 21(1). It means that a person cannot transfer ownership or title in something which he does not himself own – cannot pass on good title if he does not have it himself.

There are exceptions to the *nemo dat* rule. Where the contract is one of sale of goods or materials, the most significant exception is that created by the Sale of Goods Act 1979 section 25, which provides for a buyer in possession to transfer possession with the consent of the owner following an agreement to sell, even though the buyer does not own the goods. In such circumstances, provided that the third party purchasing from the buyer has no notice of the absence of the title in the first buyer, he, the second buyer, will obtain a good title even against the true owner.

In other words, if a contractor is paid for materials by an employer who is ignorant of the retention clause in the supplier's contract with the contractor, the employer can acquire a good title even though the supplier has not been paid.

11.12.3 In the case *Aluminium Industrie Vaassen bv* v. *Romalpa Aluminium* (1976) it was established that a seller who supplies goods under retention of title and authorises his buyer to sell them, on condition that he accounts for the proceeds of sale, has an equitable right to trace those proceeds and to recover them from the buyer.

11.12.4 An example of an effective reservation of title clause in the construction industry is to be found in the case of *W. Hanson (Harrow) Ltd* v. *Rapid Civil Engineering Ltd and Usborne Developments Ltd* (1987).

In this case, Hanson were suppliers of timber and timber products, Usborne was a development company engaged in developing some residential sites in London and the building contractor was Rapid. Hanson brought a claim against Usborne alleging the wrongful use of building materials supplied by Hanson to Rapid. Hanson had been suppliers to Rapid since 1979 and their terms of business had always been the same. They were set out on all of their documentation, i.e. consignment notes, delivery notes and invoices. One of those terms, namely condition 10, dealt with the retention of title in the following form:

> '10 Transfer of Property
> a. The property in the goods shall not pass to you until payment in full of the price to us;
> b. The above condition may be waived at our discretion where goods or any part of them have been incorporated in building or constructional works.'

Rapid and Hanson carried on business on a running account. Rapid were apparently slow payers. In order to assist Rapid with their cash flow,

Usborne had decided to make more frequent payments against the contract price by including valuations of goods on site as well as for work done.

On 16 August 1984 Hanson made a delivery to one of the residential sites. Shortly afterwards they found out that Rapid had gone into receivership on 15 August. They demanded payment or return of the goods but both were refused by the receivers, although the receivers permitted them to enter the site on 17 August to make inventories of goods for which payment had not been made. Hanson marked the goods so as to identify them.

Hanson reminded the receivers of their contract with Rapid and in particular the retention of title provision and enclosed lists of the goods stating their intention to collect them. Hanson also notified Usborne of their claim to retention of title to the goods and reserved their right to claim damages against Usborne for conversion if they used the goods. They failed to get an assurance from Usborne that Usborne would not use the goods and accordingly proceedings were commenced.

The question to be decided by the court, therefore, was whether the title to the goods remained in Hanson (the supplier) or was Usborne (the developer) protected by section 25 of the Sale of Goods Act 1979. The judge held firstly that there was no delivery or transfer of the goods on site by Rapid to Usborne by way of any sale or other disposition under section 25. The contract between Rapid and Usborne provided for monthly payments of 97% of the value of the work executed, including the value of all materials on site. The contract further provided that the property in the goods supplied should not pass to Usborne until payment of the instalment in which the supply was contained.

The judge referred to section 2 of the Sale of Goods Act 1979 which provides, *inter alia*, as follows:

> 'A contract of sale of goods is one by which the seller transfers or agrees to transfer property in the goods;
>
> Where under such a contract of sale the property in the goods is transferred from the seller to the buyer, the contract is called a sale;
>
> Where under such a contract of sale the transfer of the property in the goods is to take place at a future time or subject to some condition to be fulfilled, then the contract is called an agreement to sell.'

The judge made the point that section 25 only applies to delivery or transfer of goods under a 'sale or otherwise disposition' and that an agreement to sell only becomes a sale when any conditions are fulfilled subject to which the property in the goods is to be transferred. As between Rapid and Usborne, therefore, the building contract between them only operated to create a sale within the meaning of section 25 when any conditions subject to which the property in the goods was to be transferred were fulfilled. In this case, it meant that payment of the valuation in which the supply of materials was contained was required before any sale took place. Until then there was only an agreement to sell but not a sale. Accordingly, Usborne did not obtain any title to the goods and Hanson's claim to title was not defeated by the operation of section 25.

Hanson had effectively retained title to the goods as permitted by section

19 of the Sale of Goods Act and the title in those goods did not pass to Usborne. In so far as Usborne used them they had wrongly converted them to their own use.

If, in this case, Usborne had paid Rapid for the goods delivered to site, then the agreement to sell would have been converted into a sale. Provided Usborne were unaware of the retention of title clause between Rapid and Hanson, Usborne would have obtained a good title to the goods under section 25. This principle is shown in the Scottish case detailed below.

11.12.5 Another case also involving a judicial interpretation of section 25 of the Sale of Goods Act 1979 is *Archivent Sales & Development Ltd* v. *Strathclyde Regional Council* (1984), where Archivent were sellers and agreed to sell ventilators to the contractors for incorporation into a primary school. The contractors had a contract with Strathclyde based on JCT 63.

The Archivent conditions of supply provided as follows:

> 'Until payment of the price in full is received by the company, the property in the goods supplied by the company shall not pass to the customer.'

Ventilators were delivered to site and their value included in an interim certification under the main contract. The employer paid the contractor but the contractor failed to pay Archivent before going into receivership.

Archivent sued the employer for the value of the ventilators, arguing that title to the property remained in them. The employer contended that section 25 operated to give them an unimpeachable title. Archivant challenged this, saying that its operation depended upon a transfer of the goods from the contractor to the employer. The goods in this case were at all times under the control of the employer (not the contractor) by virtue of the provisions in clause 14(1) of JCT 63 requiring the architect's consent to any removal after they were delivered to site by the subcontractor. Archivent contended, therefore, that there could be no transfer by the contractor to the employer so as to qualify for the protection afforded by section 25.

It was held that section 25 conferred good title upon the employer. The ventilators were in the contractor's possession in law even if not under their direct control. They were transferred to the employer's possession upon the employer's surveyor making provision for payment for them in a payment certificate.

11.12.6 If the clause does not fully reserve legal title, it will be merely a charge which will be ineffective and void as against a liquidator etc., unless registered under the Companies Act 1985.

If the goods lose their identity by being completely submerged with other goods so as to produce a new material, the chances are that the retention of title clause will cease to operate.

11.12.7 Where the goods or materials are simply supplied by a subcontractor as part of their work there will be no sale of goods either to the main contractor or to the employer. The goods will pass either to the main contractor or to the Employer under a contract for work and materials. This being so, section 25 of the Sale of Goods Act 1979 will not apply unless the contractual machinery is such that the goods or materials are separately sold apart from the work element. Here, a straightforward reservation of title clause, i.e. the subcontractor retaining title in the goods or materials until he is paid, could well be effective until the goods become incorporated into the building.

However where standard forms of JCT contract and subcontract are used, this is no longer possible. JCT 80 was amended as was NSC/4 and DOM/1 as a result of the decision in *Dawber Williamson Roofing Ltd* v. *Humberside County Council* (1979) where the main contract was based on JCT 63.

The domestic subcontractor's contract was based on the standard form of domestic subcontract, referred to as the Blue Form. By clause 14 of the main contract, it was stated that any unfixed materials or goods delivered to and placed on or into the works, should not be removed without consent, and that when the value of those goods had been included in a certificate under which the contractor had received payment, such materials or goods should become the employer's property. By clause 1 of the domestic subcontract the subcontractor was deemed to have notice of all the provisions of the main contract apart from prices. It should be noted that there was no express provision in the subcontract concerning when and if the property in the subcontractor's materials or goods was to pass to the main contractor.

The subcontractor delivered 16 tons of roof slates to the site and submitted invoices to the main contractor. An interim certificate, which included the value of the slates, was issued under the main contract. The employer paid the appropriate sum to the main contractor. Therefore ownership in the slates would vest in the employer, according to the main contract, as the amount had been certified and paid. The main contractor did not pay the domestic subcontractor and went into liquidation. The subcontractor claimed that he was still the owner of the slates and therefore was entitled to their possession. The court held that the slates were still owned by the subcontractor. They were never at any time owned by the main contractor, and therefore he could not pass title in them to the employer. The employer had to pay for them again.

11.12.8 NSC/C and DOM/1 now provide that, where the value of any materials or goods has been included in any interim certificate under which the amount properly due to the contractor has been paid for by the employer, the materials or goods become the property of the employer and the subcontractor cannot deny that this is so. There is further provision in JCT 98 requiring the main contractor to insert a similar provision in any subcontract issued. Similar terms are incorporated into IFC 1998.

Even with the amendments, it is still quite possible that the employer's position will not be secure. This will happen when either the contractor uses his own form of subcontract or perhaps, more often, when the supplier to the subcontractor has himself got a reservation of title clause.

11.12.9 Once goods or materials become incorporated into the works the effectiveness of a retention of title clause will be lost. In the case of *Peoples Park Chinatown Development Pte Ltd* v. *Schindler Lifts (Singapore) Pte Ltd* (1993) Schindler was appointed as a nominated subcontractor to the main contractor. The developer was Peoples Park Chinatown who became insolvent before the project was completed. Before Schindler was nominated it entered into an agreement with Peoples Park Chinatown to accept deferred payment terms. Schindler supplied and installed ten escalators but did not test or commission them nor supply and fix the finishings such as balustrades. The liquidator of Peoples Park Chinatown sold the building together with the ten escalators to a third party. Schindler claimed against the liquidator for payment for the ten escalators. The liquidator refused on the grounds that the escalators had become attached to the building (i.e. the land) and therefore belonged to Peoples Park Chinatown. He claimed a right to sell them leaving Schindler as an unsecured creditor.

It was held:

(1)   The escalators had been fixed to the building in such a way as to become a permanent feature of the building. They became part of the land notwithstanding they were not commissioned. Therefore they became the property of Peoples Park.

(2)   There was no direct agreement between Schindler and Peoples Park. The escalators had been supplied as part of the main contract between Peoples Park and the main contractor. There was no provision in the main contract that would give Schindler the right to recover the escalators from the building after they were fixed. However, they could have recovered any unfixed materials (e.g. the balustrades).

When this case was heard at first instance the judge found in favour of Schindler which seemed just in the light of the terms of payment agreed between Schindler and Peoples Park. However, the Court of Appeal reverted to the classical position in English land law. When items become permanently fixed to a building they become part of the land and they belong to the owner of the land. The court followed the principal English authority on this point: *Seath & Co* v. *Moore* (1886).

11.12.10 The Insolvency Act 1986 imposes certain restrictions on retention of title clauses. Once a petition has been presented to the court for the appointment of an administrator to a company section 10(1)(b) of the 1986 Act provides that no steps may be taken to repossess goods in the company's possession under any retention of title agreement except with the leave of the court and subject to such terms as the court may impose. Once an administrator has been appointed the embargo on repossession continues, though an administrator as well as the court may consent to the recovery of goods by the supplier.

Moreover an administrator is empowered to sell goods which have been supplied subject to a retention of title agreement if he can persuade the court that disposal would be likely to promote one or more of the purposes specified in the administration order.

SUMMARY

Reservation of title clauses, if properly drafted, will normally provide protection to a supplier who delivers goods to a purchaser who becomes insolvent before payment is made. However, no protection exists if the goods have been incorporated into the works.

A further difficulty can arise for the seller under section 25 of the Sale of Goods Act 1979. Where this section applies, a purchaser acting in good faith without notice of the reservation of title may acquire a good title despite the provisions of the clause.

The Insolvency Act 1986 provides that, where a petition has been presented to the court for the appointment of an administrator goods cannot be repossessed under a reservation of title clause except with the leave of the court.

11.13 Can the signing of time sheets which make reference to standard conditions of contract form the basis of a contract?

11.13.1 The courts seem ever willing to infer that a contract has come into being in preference to the uncertainty where goods are bought and sold or work carried out in the absence of a contract. Agreement between the parties is one of the main ingredients in deciding whether there is a contract. In trying to establish whether agreement has taken place courts are prepared to examine the business dealings of the parties.

11.13.2 This matter was the subject of the decision in *Grogan* v. *Robin Meredith Plant Hire and Triact Civil Engineering* (1996).

In 1992 Triact, a civil engineering contractor, was laying pipes on a site. The principal of Meredith, a plant hire company, approached the site agent and it was agreed that Triact would hire a driver and machine from Meredith at an all in rate of £14.50 per hour from 27 January 1992. No formal agreement was mentioned. At the end of the first and second weeks, Meredith's driver presented the site agent with time sheets for checking and signature. At the bottom of the sheet it stated: 'All hire undertaken under CPA conditions. Copies available on request.'

During the third week, there was an accident and the plaintiff was injured. He issued proceedings against both defendants and they consented to judgment in the sum of £82 798.17, Meredith paying one third and Triact two thirds. Meredith claimed that Triact was liable to indemnify it in accordance with the CPA conditions referred to on the time sheets. The issue was whether the presentation of the time sheets by the driver amounted to a variation of the plant hire contract sufficient to incorporate the CPA conditions by reference.

The central question was whether the time sheet had a contractual effect. It was held that normally a document such as a time sheet, invoice or statement of account does not have a contractual effect in the sense of making or varying a contract. Time sheets do not normally contain evidence as to the terms of a contract, and in this case they were intended merely as a record of a party's performance of an existing obligation. The signed time sheets did not have, nor purport to have, contractual effect. Therefore Triact was not liable to indemnify Meredith.

SUMMARY

> If time sheets which make reference to standard conditions of contract are signed it does not follow that those conditions will form part of the contract or become substituted for existing conditions of contract.

11.14 Can suppliers rely upon exclusion clauses in their terms of trading to avoid claims for supplying defective goods or claims based on late supply?

11.14.1 Suppliers to the construction industry, when drafting their conditions of trading, will usually include a clause the effect of which will be to reduce or even in some cases eliminate their liability to reimburse a purchaser where goods are delivered late or are in some way defective.

In an effort to regulate this type of limitation or exclusions clause the Unfair Contract Terms Act 1977 was enacted. The statute is unfortunately worded as it deals only with clauses which seek to limit or exclude liability and not unfair terms in general.

11.14.2 Section 3 of the Act states that it applies to contracts made after 1 February 1978 and relates to all clauses excluding or restricting liability which are contained in 'written standard terms of business'. A contracting party who wishes to rely on any such term must demonstrate to the satisfaction of the judge (or arbitrator) that the exclusion or limitation is something which it was 'fair and reasonable' to have included in the contract 'having regard to the circumstances which were, or ought reasonably to have been, known to or in the contemplation of the parties when the contract was made'.

There have been a number of cases related to the construction industry where courts have been called upon to decide whether an exclusion or limitation clause is valid.

11.14.3 In the case of *Rees Hough Ltd* v. *Redland Reinforced Plastics Ltd* (1984) the court had to decide whether the following limitations clause which formed part of Redland's standard terms of trading was reasonable and therefore enforceable.

> 'The company warrants that the goods shall be of sound workmanship and materials and in the event of a defect in any goods being notified to the company in writing immediately upon the discovery thereof which is the result of unsound workmanship or materials, the company will, at its own cost at its option, either repair or replace the same, provided always that the company shall be liable only in respect of defects notified within three months of delivery of the goods concerned. Save as aforesaid, the company undertakes no liability, contractual or tortious, in respect of loss or damage suffered by the customer as a result of any defect in the goods (even if attributable to unsound workmanship or materials) or as a result of any warranty, representation conduct or negligence of the company, its directors, employees or agents, and all terms of any nature, express or implied, statutory or otherwise, as to correspondence with any particular description or sample, fitness for purpose or merchantability are hereby excluded.'

In deciding that the limitation clause was unfair and therefore unenforceable, the court took the following matters into account:

- The strength of the bargaining positions of the two parties
- Whether the customer received an inducement to agree to the term or had
  an opportunity of entering into a similar contract with others without such
  a term
- Whether the customer knew of the term
- Where the contract excluded or restricted liability for breach of condition,
  whether it was reasonable to expect compliance with it;
- Whether the goods were manufactured to the special order of the
  customer.

Section 11(5) of the Act provides that: 'It is for those claiming that a contract
term ... satisfies the requirement of reasonableness to show that it does.'

11.14.4 In *Chester Grosvenor Hotel Co Ltd* v. *Alfred McAlpine Management Ltd* (1991) a
dispute arose concerning whether an exclusion clause in a management
contract was reasonable.

The wording stated that McAlpine would undertake to take all practical
steps to enforce the construction contractors' contracts, to secure performance of the obligations under those contracts and to recover damages. Such
action was to be in Grosvenor's name and at Grosvenor's expense. The
clause also said that Grosvenor were not entitled to recover from McAlpine,
by set-off or other action, any sums greater than those which McAlpine
recovered with Grosvenor's consent from the construction contractors.
Further, Grosvenor could not recover any such sums from McAlpine before
McAlpine themselves had recovered them from the construction
contractors.

Grosvenor argued that the exclusion clause was subject to and invalidated
by section 3 of the Unfair Contract Terms Act 1977. The judge in finding the
clause as reasonable identified the following factors in favour of McAlpine:

- The equal bargaining power of the parties
- The fact that the clause was designed to act as an agreed division of risk
  between commercial entities dealing at arm's length
- The fact that this allocation of risk was reflected in McAlpine's remuneration
- The relatively slight risk to which Grosvenor were exposed under the
  clause (viz. that of McAlpine's insolvency, it being plain that Grosvenor
  still had a right to proceed directly against the construction contractors
  under their direct contracts)
- Grosvenor's right to control that risk by vetoing the selection of construction contractors under clause 1(c) and 19 of the management contract
  and by making inquiries about those contractors' financial status
- The fact that it was within the parties' contemplation that, if the risk to
  Grosvenor eventuated, the results would probably not be disastrous
- The fact that McAlpine's contracts had been presented to Grosvenor as
  open to negotiation and not on a 'take it or leave it' basis

- The fact that Grosvenor could have contracted on different terms with other management contractors
- The substantial time and general opportunities available to Grosvenor for consideration of the terms
- The availability to Grosvenor of independent advice
- The intelligibility of the exclusion clause
- The fact either party could have insured against the relevant risk, but that in either event the cost would have fallen on Grosvenor.

11.14.5 Other cases involving limitation or exclusion clauses relating to construction work include the following:

- *Charlotte Thirty and Bison* v. *Croker* (1990)
  In this case it was held that an exclusion clause in a contract to supply batching plant was unreasonable and therefore unenforceable.

- *Barnard* v. *Marston* (1991)
  In this case a subcontractor's exclusion clause was held to be valid.

- *Barnard Pipelines Technology* v. *Marston Construction* (1992)
  An exclusion clause in this case was held to be reasonable and therefore valid. The fact that the purchaser was aware of the exclusion clause when the contract was entered into influenced the judge.

- *Stewart Gill* v. *Horatio Myer* (1992)
  In this case a right of set-off was excluded by the terms of trading. The court held that this was unreasonable and therefore invalid.

11.14.6 In the case of *British Fermentation Products Ltd* v. *Compair Reavall Ltd* (1999) the court had to check whether terms and conditions which applied to a contract were one party's 'standard terms of business' as referred to in the Act. It was held that for this description to apply the term must 'invariably or at least usually be used by the party in question'.

11.14.7 It is not uncommon for a supplier's terms and conditions to include a clause which requires the purchaser to inspect the goods on delivery. The clause usually goes on to either exclude or limit liability for defects which could have been discovered at the inspection stage.

In the case of *Expo Fabrics (UK) Ltd* v. *Naughty Clothing Co Ltd* (2003), the Court of Appeal had to decide whether an exclusion clause in a contract to supply fabrics passed the reasonableness test. Clause 6A in the terms stipulated that any claim based upon a defect in the quality or condition of the goods should be notified in writing within 20 days of delivery. The clause went on to say that outside this period Expo would have no liability for faulty goods. Clause 6B dealt with the situation where faults were discovered in the goods. When this occurred Expo was obliged to replace the goods free of charge or refund the price.

Lord Justice Waller, in the Court of Appeal, considered the clause reasonable as the 20-day period provided plenty of time to check the goods.

11.14.8 Disputes as to whether exclusion or limitation clauses are reasonable are a constant source of litigation as can be seen from the following cases:

> *The Salvage Association* v. *CAP Financial Services* (1992)
> *Trolex Products* v. *Merrol Protection Engineering* (1991)
> *Fillite Runcorn* v. *APV Plastics* (1993)
> *Edmund Murray* v. *BSP International Foundations* (1992)
> *St Albans City & District Council* v. *International Computer Ltd* (1996)
> *Omega Trust* v. *Wright Son and Pepper* (1996)
> *Governor & Company of the Bank of Scotland* v. *Fuller Peiser* (2001)
> *Mostcash plc (Formerly UK Paper), Fletcher Challenge Forest Industries Ltd*
>   *and Metsa-Serla (Holdings) Ltd* v. *Fluor Ltd* (2002)
> *John Moodie & Co and Others* v. *Coastal Marine (Boatbuilders) Ltd* (2002)
> *Bacardi-Martini Beverages Ltd* v. *Thomas Hardy Packaging* (2002)
> *Britvic Soft Drinks Ltd and Others* v. *Messer UK Ltd and Another* (2002)
> *Amiri Flight Authority* v. *BAe Systems plc and Crossair Limited Company for*
>   *Regional European Air Transport* (2002)
> *SAM Business Systems Ltd* v. *Hedley & Co (A Firm)* (2002)
> *Rolls Royce Engineering plc and Another* v. *Ricardo Consulting*
>   *Engineers Ltd* (2003)
> *Farrans Construction Ltd* v. *RMC Ready Mixed Concrete*
>   *(Scotland) Ltd* (2003)
> *University of Keele* v. *Price Waterhouse* (2004)

11.14.9 The unfortunate aspect of this statute is that the courts, whilst trying to be consistent in the way in which they have decided when limitation or exclusion clauses are unreasonable and therefore unenforceable, have left suppliers with a dilemma. When entering into contracts to supply their products they do not know whether the exclusion or limitation clauses will be valid should they have need to make use of it. This being the situation, how do they price the risk?

SUMMARY

Suppliers are entitled to include in their terms of trading, clauses which exclude or limit their liability if goods they supply prove to be defective or are delivered late. However under the Unfair Contract Terms Act 1977 if such clauses are to be enforceable the supplier is required to demonstrate that they are reasonable.

11.15 What level of supervision must an architect provide on site?

11.15.1 Where work proves to be defective and the contractor becomes insolvent and unable to correct the defects questions are often asked as to whether the architect has any liability to the employer for the cost of correcting the defects. This raises the matter as to the extent to which the architect is obliged to supervise the work of the contractor. A typical clause used by architects when drafting their own terms is:

> 'At intervals appropriate to the stage of construction visit the works to inspect the progress and quality of the works and determine that they are being executed generally in accordance with the contract document.'

11.15.2 The matter of the obligation of the architect with regard to supervision has been the subject of a number of cases. In *Alexander Corfield* v. *David Grant* (1992) the plaintiff architect sought unpaid fees of £23 750.95 plus interest for work undertaken at a listed building whose use the defendants wished to change to a hotel. The defendants counterclaimed for damages for breach of contract.

The defendants had a set time scale for the conversion work in order to get the hotel into the 1990 guide books which were to go to press in May 1989. In a letter, the plaintiff set out a proposed timetable and recommended that the RIBA fee scale be used. The defendant wrote back accepting this and agreeing that plans must be submitted from time to time to meet the planning requirements and stressed that this should be done in time for the district council planning meeting. The defendant also accepted the plaintiff's recommendation as to the builder to be employed.

The defendant submitted that the plaintiff made so many mistakes that he was in repudiatory breach of the contract. The court accepted that the plaintiff continually did or omitted things to such an extent that his conduct showed an intention not to perform the contract to a reasonable standard, and that the defendant reasonably lost confidence in him so as to entitle them to accept the breaches as repudiation of the contract and to dismiss him.

The court in arriving at a decision advised on the level of supervision on site which an Architect is required to provide. What was adequate by way of supervision and other work was not in the end to be determined by the number of hours worked, but by asking whether it was enough for the job. In this case the plaintiff needed, but did not have, at least one skilled and experienced assistant with the result that the project was an inadequately controlled muddle and the plaintiff was in continuous breach of contract.

Judge Stabb held

> I think that the degree of supervision required of an architect must be governed to some extent by his confidence in the contractor. If and when something occurs which should indicate to him a lack of competence in the contractor, then, in the interest of his employer, the standard of his supervision should be higher. No one suggests that the architect is required to tell a contractor how his work is to be done, nor is the architect responsible for the manner in which the contractor does the work. What his supervisory duty does require of him is to follow the progress of the work and to take steps to see that those works comply with the general requirements of the contract in specification and quality. If he should fail to exercise his professional care and skill in this respect he would be liable to his employer for any damage attributable to that failure.

11.15.3 Judge Stabb reflected what Lord Upjohn said about the nature of an architect's duty in circumstances where he knew about the incompetence of the contractor. This was In *East Ham Corporation* v. *Bernard Sunley & Sons Ltd* (1964) where he said with regard to supervision by an architect:

> As is well known, the architect is not permanently on the site but appears at intervals, it may be a week or a fortnight, and he has, of course, to inspect the progress of the work. When he arrives on the site there may be very many important matters with which he has to deal: the work may be getting behind-hand through labour troubles; some of the suppliers of materials or the subcontractors may be lagging; there may be physical trouble on the site itself, such as finding an unexpected amount of underground water. All these are matters which may call for important decisions by the architect. He may in such circumstances think that he knows the builder sufficiently well and can rely upon him to carry out a good job; that it is more important that he should deal with urgent matters on the site than that he should make a minute inspection on the site to see that the builder is complying with the specifications laid down by him . . . It by no means follows that, in failing to discover a defect which a reasonable examination would have disclosed, in fact the architect was necessarily thereby in breach of his duty to the building owner so as to be liable in an action for negligence. It may well be that the omission of the architect to find the defect was due to no more than error of judgment, or was a deliberately calculated risk which, in all the circumstances of the case, was reasonable and proper.

11.15.4 In *Sim and Associates* v. *Alfred Tan* (1997), a Singapore case, Alfred Tan bought a piece of land on which to build a two storey bungalow and engaged Sim & Associates, a firm of architects, planners and engineers, to provide the various services in connection with the construction of the bungalow. Sim and Associates, on behalf of Alfred Tan, awarded the contract for the building project to Hok Mee Construction.

Alfred Tan sued Sim and Associates alleging that they had breached their duties as the architect by certifying defective works and not calling upon the main contractor to rectify the defective works and complete the uncompleted works. Sim and Associates counterclaimed for the balance of their professional fees and disbursements.

The trial judge, although he rejected both grounds of claim (partly for lack of evidence and partly because the contractors refused to return because of their disagreements with Tan over various issues), nevertheless went on to hold that the building defects were attributable to a lack of proper supervision and a failure to require the contractor to make good defective works.

Sim and Associates appealed.

It was held that an architect is merely required to give the building works reasonable supervision.

(1) The defective works were ultimately the result of the failure of Tan and the main contractor to arrive at an acceptable compromise and not in any way attributable to any failure by Sim and Associates to perform their duties properly as architects.

(2) Tan had not succeeded in showing that he had suffered any real loss. The total cost of the rectification works carried out did not exceed the final sum due to the architects.

The mere existence of defective works does not, of itself, translate into a finding of lack of supervision against the architect in a building contract.

11.15.5 In *Department of Heritage* v. *Stevenson Varming Mulcahy and Balfour Beatty* (1998) many thousands of metres of cable were damaged and condemned. The court held that the damage was caused by bad workmanship. An action was brought against both the contractor who carried out the work and the engineer for failing to properly supervise.

The court held that the contractor was totally liable and the damage was caused when drawing the cables. No liability was attached to the engineer as the court considered that the engineer could not be expected to be aware that cables were being damaged when drawn. Only the operative would be aware of the damage.

SUMMARY

The architect's duties with regard to what level of supervision he must provide are rarely given in any detail in the architect's conditions of appointment. In the final analysis the court will make a subjective judgment based upon the facts of each case. However, as a general rule, an architect is merely required to provide a reasonable level of supervision. The fact that work is certified which turns out to be defective does not in itself mean that the architect has been negligent.

11.16 Where a specification includes a named supplier 'or other approved', can the architect/engineer refuse without good reason to approve an alternative supplier proposed by the contractor/subcontractor?

11.16.1 The question of the architect's right to refuse to give approval to an alternative supplier proposed by the contractor where the contract documents named a supplier 'or other approved' was the subject a dispute in *Leedsford Ltd* v. *The City of Bradford* (1956). The defendant council wished to have a new infants' school built. Contract documents including bills of quantities were prepared and submitted to contractors for pricing. Provision was made in the bills for artificial stone which was to be used in the following terms:

> 'Artificial Stone ... The following to be obtained from the Empire Stone Company Limited, 326 Deansgate, Manchester or other approved firm...'

11.16.2 The successful contractor proposed that the artificial stone should be supplied by HK White (Precast Concrete Works) Ltd and Spencer Parkinson and Sons Ltd and sought the architect's approval. The cost of stone from Empire Stone was £1250 compared with £500 from the alternative supplier.

The architect refused to approve the alternative supplier insisting the stone be obtained from Empire Stone.

11.16.3 The matter was referred to the court by the contractor who claimed the difference in the cost of stone. The contractor contended that those words had been inserted for its benefit. It was argued that they had the right to put forward and obtain approval for any firm who would supply stone of the proper quality at a price less than the price which the Empire Stone Company Ltd would charge. It was further argued that the architect had broken the contract when he insisted that the stone was obtained only from the Empire Stone Company Ltd.

11.16.4 It was held in the Court of Appeal that the words 'or other approved firm' did not give the contractor an option to submit any firm of their choice for the architect's approval. The words 'to be obtained from the Empire Stone Company Limited … or other approved firm' should be analysed in the following way:

> The builder agrees to supply artificial stone. The stone is to be Empire Stone unless the parties both agree some other stone, and no other stone can be substituted except by mutual agreement. The builder fulfils his contract if he provides Empire Stone, whether the Bradford Corporation want it or not; and the architect can say that he will approve of no other stone except the Empire Stone.

Accordingly, the position was that there was an absolute obligation on the contractor to supply Empire Stone unless the architect should give his approval to some other stone. Moreover, the architect was not bound to give any reasons for withholding approval of any other firm. The most that was required of the architect was that he should act in good faith and no allegation against his good faith was made.

11.16.5 This decision would not now apply on contracts let in the public sector where there is an EU restriction on the use of a single specified supplier.

SUMMARY

Where a contract calls for good to be supplied by a named supplier 'or other approved' the architect may refuse to approve an alternative supplier proposed by the contractor without giving reasons. The architect is only required to act in good faith.

This does not apply to public contracts where under EU rules there is a restriction on the naming of a sole supplier in a specification.

11.17 If a subcontractor is falling behind programme and in danger of completing late due to his own inefficiencies, can the contractor bring other labour on to the site to supplement the subcontractor's efforts to ensure completion on time?

11.17.1 Late completion of a project will usually result in substantial extra costs being incurred by the contractor. Where delays result from poor progress on the part of a particular subcontractor, the main contractor will usually apply pressure on the subcontractor to increase productivity to ensure that completion on time is achieved. Frustration on the part of the contractor often sets in if there is no obvious improvement in progress. Contractors' minds may then turn to the idea of bringing on to the site more operatives to supplement those of the defaulting subcontractor in an effort to make up lost time. Subcontractors are understandably resistant to operatives who are working for either the contractor or other subcontractors being employed on their project, and often a stalemate occurs or the contractor takes matters into his own hands and goes ahead and recruits more operatives. Is the contractor entitled to act in this fashion or is he subject to some form of legal restriction?

11.17.2 This matter was the subject of a dispute which came before the courts in the case of *Sweatfield* v. *Hathaway Roofing* (1997). Sweatfield, who was formally IT Design and Build, subcontracted the roofing and cladding of a leisure centre to Hathaway Roofing. The relationship between the parties was strained and not improved due to Hathaway Roofing's work falling behind schedule. A payment which was due to Hathaway Roofing was withheld by IT Design and Build and as a result Hathaway Roofing withdrew its labour from site. Negotiations between the parties resulted in Hathaway Roofing returning to the site.

Progress did not improve very much, resulting in a letter being sent by IT Design and Build to Hathaway Roofing accusing them of failing to proceed regularly and diligently with the works. Hathaway Roofing's response was to allege that their progress was being hampered by the poor workmanship of IT Design and Build. Hathaway Roofing was then instructed by IT Design and Build to increase the numbers of on-site operatives. Hathaway Roofing refused on the grounds that it was impractical to accommodate any more operatives. IT Design and Build decided to take matters into their own hands and brought on to the site a gang of sheeters from another subcontracting company. Hathaway Roofing's reaction was to withdraw its labour from the site claiming that in bringing another subcontractor on to the site to carry out some of its work, the contractor was in breach of contract. The court decided that the action of IT Design and Build was of great gravity and concluded that it was in breach of contract.

11.17.3 It is possible to write into a contract a clause which gives the contractor a right to bring on to site operatives to supplement those employed by the subcontractor.

SUMMARY

A contractor who brings operatives on to the site or introduces another subcontractor to supplement the work force employed by one of its subcontractors, risks a claim being levied for breach of contract. A clause, however, can be drafted into the subcontract which gives the contractor express power to bring on extra labour. If a subcontractor is failing to proceed regularly and diligently with the subcontract work, there is usually an entitlement under the terms of the subcontract for the contractor to terminate the subcontract.

11.18 Where a contract requires the parties to act in good faith, is it enforceable?

11.18.1 The UK government and others who regularly commission construction work have abandoned the appointment of contractors on the basis of lowest price and adopted best value as a replacement. It is now quite common to see clauses written into contracts which require the parties to act toward each other in good faith. The 1998 version of GC/Works/1, for example, includes the following wording:

> 'The Employer and the Contractor shall deal fairly in good faith and in mutual co-operation with one another and the Contractor shall deal fairly in good faith and in mutual co-operation with all his subcontractors.'

There is little in English law which is of assistance in defining what is meant by good faith. The GC/Works contract provides helpful provisions which differ from the normal contract requirement and relate to actions of the parties which amount to expressions of good faith. There is, for example, a requirement for the parties to hold project team meetings. At these meetings the parties will jointly develop proposals for reducing costs and give advance warning of any matters which may lead to an increase in costs or affect the final quality of the work.

11.18.2 The case of *Cable & Wireless plc* v. *IBM United Kingdom Ltd* (2002) arose out of a dispute concerning a contract in which there was a clause obliging the parties to act in good faith. The parties entered into a global framework agreement for the supply of information technology services. Clause 41 set out the following obligations:

> (1) 'The parties shall attempt in good faith to resolve any dispute or claim arising out of or in relation to this agreement... promptly through negotiations between the respective senior executives of the parties who have authority to settle the dispute pursuant to clause 40.'
> (2) 'If the matter is not resolved through negotiation, the parties shall attempt in good faith to resolve the dispute or claim through an alternative dispute resolution (ADR) procedure ...'

The claimant argued that this procedure was only an agreement to negotiate and not binding. It was held, however, that the reference to ADR was analogous to an agreement to arbitrate and represented a free-standing agreement which was ancillary to the main contract. As such it was capable of being enforced by a stay of proceedings or by injunction. The parties were therefore required to seek a settlement of the dispute by way of negotiations and if that failed to undertake an ADR procedure.

11.18.3 In the absence of an express clause which requires the parties to act in good faith, the courts may in the final analysis decide that there is a requirement to act in good faith implied in the contract. In the case of *Balfour Beatty* v. *Docklands Light Railway* (1996), the Court of Appeal held that the employer was required to act honestly, fairly and reasonably. The implied requirement for a party to draw attention to the onerous nature of clauses in its contract was recognised by the court in the case of *Thornton* v. *Shoe Lane Parking Ltd*

(1971) and is an example of a requirement to act in good faith. In the case of *Interfoto Picture Library Ltd* v. *Stilleto Visual Programmes Ltd* (1989) the judge said of good faith:

> 'that it does not simply mean [the parties] should not deceive each other ... its effect is perhaps most aptly conveyed by such metaphorical colloquialisms as playing fair, coming clear or putting one's cards on the table. It is in essence a principle of fair and open dealing.'

There are other examples of acts of good faith which have been recognised by the courts. In the cases of *Blackpool & Fylde Aero Club* v. *Blackpool Borough Council* (1990) the court held that a tenderer, if he complies with the requirements of the tender enquiry, is, as a matter of contractual right, entitled to have his tender considered along with the others which have been submitted.

11.18.4 A failure to act in good faith, whether the requirement is an express or implied term in the contract, will entitle the injured party to recover the costs incurred as a result of the failure. In some cases the loss arising from the breach can be fairly established in principle, for example failing to consider a tender which has been properly submitted, but the sum involved may be a little more difficult to ascertain with any accuracy. The same may be said of a party who regularly refuses or neglects to attend a team meeting to try and identify cost savings. The courts, however, have never been slow at being inventive when it comes to ascertaining loss which in principle is recognised but at first glance is difficult to establish with accuracy. If it is impossible to demonstrate loss due to a failure to act in good faith, then there will be no case to answer.

SUMMARY

There is a growing tendency to include in contracts a requirement that the parties are to act in good faith. Where there is no express provision, the courts may decide that the requirement of good faith is implied. There is a need to be able to define what is meant by good faith and it is helpful if the wording in the contract is expressive of the meaning. Where a breach of good faith occurs, the injured party will be entitled to recover the costs which have been incurred as a result of the breach. It may often be the case that, whilst the principle of loss can be easily established, its assessment with anything like accuracy becomes a difficult matter. The courts, however, are inventive in their approach to deciding on an amount where this occurs.

11.19 Can parties to a dispute be forced to submit the matter to mediation?

11.19.1 ADR has developed into a fast and cost effective means of dispute resolution. Mediation has, for a number of years, been accepted as one of the most popular of the various ADR techniques. Some conditions of contract include mediation as part of the disputes resolution procedures.

11.19.2 In the case of *Cable and Wireless plc* v. *IBM United Kingdom Ltd* (2002) (see section 11.18.2), the court had to decide whether, under the terms of a contract to supply information technology, the parties could be forced to refer the matter for resolution by ADR in accordance with clause 41. The court held that the parties were obliged to submit to ADR.

11.19.3 Under the Civil Procedures Rules, High Court judges have power to recommend parties who are involved in litigation to submit their dispute to mediation.

11.19.4 Courts are prepared to penalise parties who refuse mediation without justifiable grounds. In the case of *Dunnett* v. *Railtrack* (2002), a dispute was referred to the Court of Appeal where Railtrack were successful. The normal rule that the losing party pays the winner's costs was expected to apply. Despite Railtrack's success, the Court of Appeal refused to order that Dunnett pay Railtrack's costs as Railtrack had refused the offer of mediation.

11.19.5 The matter of a costs order was again the subject of a dispute in the case of *Hurst* v. *Leeming* (2002). Hurst was a partner in a firm of solicitors, and fell out with his partners. The matter was referred to court and Hurst instructed Ian Leeming QC to represent him. The case failed in the Lower Court, the Court of Appeal and in the House of Lords. Hurst considered that Leeming had negligently handled his case and commenced an action in the High Court. Following advice from the judge, the case was discontinued. Normally Hurst would have been left to pay Leeming's costs but after the commencement of proceedings Hurst had invited Leeming to partake in mediation, but this offer was refused. In view of the decision in the Railtrack dispute, Hurst sought to have Leeming's refusal of mediation to be taken into account. The judge, however, was persuaded that, as Hurst had no real prospect of success, this was justification for Leeming refusing to take part in mediation.

11.19.6 In *Malkins Nominees Ltd* v. *Societe Finance* (2002), the successful party lost 25% of its overall cost recovery after refusing an offer of mediation.

11.19.7 Rather surprisingly, in the case of *Shirayama Shokusan Co Ltd and Others* v. *Danova Ltd* (2003), the court decided that it had the power to force the parties to mediate even though neither wished to take this route. Mediation is very much a process of settlement when both parties are prepared to enter into meaningful negotiations. If one or both parties are looking for an outright victory, mediation could become a waste of time and cost.

11.19.8 The Court of Appeal in *Halsey* v. *Milton Keynes General Hospital Trust* (2004) laid down some guidelines as to when it would be inappropriate for mediation to be used as a method of attempting to resolve a court action:

(1) If the whole object of the litigation is to determine an issue of law or lay down a legal principle

(2) Where the claimant's position is unreasonably weak and is attempting to use mediation as tactical play in a nuisance value action

(3) Where a defendant has made a reasonable offer to settle, which has been rejected

(4) The cost of mediation would be disproportionately high

(5) If the prospect of mediation would delay the date for trial

(6) Where mediation has little chance of success due to an intransigent attitude of one or both parties.

SUMMARY

A party to a dispute which arises under a contract where the disputes resolution procedures provide for mediation, can be forced to have the dispute referred to mediation. In the absence of provision within the contract for mediation, should the matter be the subject of litigation, the judge has power to recommend that the parties refer the matter to mediation. If one of the parties refuses to submit to mediation, it may affect its entitlement to the recovery of costs in the event of it receiving a favourable court award.

12
ADJUDICATION

Contractual Problems Nos 12.1 to 12.25 arise out of adjudication under Part II of the Housing Grants, Construction and Regeneration Act 1996 referred to as the 'Construction Act'.

12.1 Will an adjudicator's award be enforced by the courts using a summary procedure?

12.1.1 Prior to an enforcement case coming before the courts, it was considered by some eminent authorities that the courts would not, and in fact could not, enforce an adjudicator's award by a summary procedure. The reasons were based upon a number of cases, including *Halki Shipping Corporation* v. *Sopex Oils Ltd* (1997).

12.1.2 The first court case concerning the enforcement of an adjudicator's decision was *Macob Civil Engineering Ltd* v. *Morrison Construction* (1999). In this case the parties entered into a contract under which the claimant Macob was to carry out groundworks. A payment dispute arose which was referred to Mr E. Mouzer, an adjudicator. His decision was that the subcontract provided a payment mechanism which did not comply with the Act, consequently the Scheme for Construction Contracts applied. Mr Mouzer concluded that the defendant had served a notice indicating an intention to set off from payments due out of time and therefore immediate payment should be made to Macob. The adjudicator's decision was in the form of a pre-emptory order under which either party could apply to the court for enforcement.

The defendant applied to the court for a stay to arbitration on the grounds that the decision was wrong on its merits and also that there had been a breach of the rules of natural justice. The judge rejected the defendant's arguments, holding that to refuse enforcement would substantially undermine the effectiveness of adjudication. He considered that the provision of the Act should be construed positively. The decision of the adjudicator was therefore held to be binding.

12.1.3 In *Outwing Construction Ltd* v. *H. Randall* (1999), a dispute under DOM/1 was referred to Mr Talbot, an adjudicator appointed by the Chartered Institute of Building. He considered that, as the terms of the subcontract did not comply with the Act, the Scheme for Construction Contracts applied.

It was the decision of Mr Talbot that the defendant should pay the claimant, plus his fees and expenses. He ordered his decision to be made pre-emptorily. The claimant issued an invoice for the amount included in the adjudicator's decision. In the absence of payment, solicitors acting for the claimant provided a deadline for payment threatening to apply for summary judgment in the event of non-payment. After the date had passed, the defendant responded to the effect that it intended to seek a court stay to arbitration. A writ was issued by the claimant for payment plus interest and costs. The defendant made payment but refused to pay the claimant's costs. The court found in favour of the claimant. It was the view of the judge that the intention of Parliament is clear: disputes may be referred to adjudication and the decision of the adjudicator has to be complied with.

12.1.4 It was not long before the Scottish court became involved with regard to adjudication. In the case of *Rentokil Allsa Environmental Ltd* v. *Eastend Civil Engineering Ltd* (1999), the adjudicator found in favour of the pursuer. A cheque for the full amount was provided by the defendant, who simultaneously lodged an arrestment freezing the payment they were obliged to make to the pursuer. This is a mechanism in Scotland which does not apply in England whereby, to put pressure on a reluctant payer, the payee seeks to arrest or freeze payments due to the payer by third parties, in this case freezing payment from itself to the pursuer. The court found in favour of the pursuer, refusing to allow the use of the arrestment mechanism to circumvent and negate the effect of the adjudicator's decision.

SUMMARY

It has been made clear by the courts that in passing the Construction Act the intention of Parliament is clear in that a decision of an adjudicator appointed under the Act can be enforced by summary procedure.

12.2 Will the court enforce part only of an adjudicator's award?

12.2.1 An adjudicator's decision may comprise a number of parts. If the court holds that part of the award is tainted and unenforceable, will the valid parts be enforced? This was the problem faced by the court in the case of *R. Durtnell and Sons* v. *Kaduna Ltd* (2003). Kaduna and Durtnell entered into a JCT 80 standard form of contract for work to be undertaken at Kaduna's property in Hampshire. The parties were in dispute concerning the sum which Durtnell was due to be paid under the terms of the contract and their entitlement to an extension of time.

12.2.2 The adjudicator made a declaration concerning Durtnell's entitlement to an extension of time and an award of £1.2m in respect of the payment claim. For some strange reason Kaduna paid half of the £1.2m and then contested the adjudicator's decision on the basis that, under the terms of the contract, the architect had 12 weeks to deal with applications for extensions of time and the period had not expired. Durtnell applied to the court to have the balance of the adjudicator's decision paid.

12.2.3 The court agreed with Kaduna that the adjudicator should not have included an entitlement to an extension of time in his decision as the 12-week period available to the architect for granting extensions of time had not elapsed. However, the court ordered Kaduna to pay to Durtnell the balance of the £1.2m.

SUMMARY

In the case of *R. Durtnell and Sons* v. *Kaduna Ltd* (2003), the adjudicator in his decision awarded £1.2m and a declaration concerning an extension of time. Unfortunately the adjudicator acted outside his jurisdiction concerning the extension of time. Nonetheless, the court ordered the payment to be made.

12.3 When can it be said that a dispute has arisen giving rise to an entitlement for the matter to be referred to adjudication?

12.3.1 Section 108 (1) of the Construction Act states that:

'A party to a construction contract has the right to refer a dispute arising under the contract for adjudication...'

12.3.2 If there is no obvious dispute there is nothing to be referred to adjudication. This may seem obvious but nonetheless, in the case of *Sindall Ltd* v. *Sollard* (2001), a matter was referred to an adjudicator before a dispute was alleged to have taken place. Sindall was the main contractor for the refurbishment of Lombard House in Mayfair. The employer was Sollard and the contract administrator Michael Edwards. Work was delayed and the contractor requested an extension of time. A period of 12 weeks was awarded which did not meet with Sindall's minimum requirements and the matter was referred to adjudication. The adjudicator decided that an extension of 28 weeks was appropriate. Further delays occurred and the employer threatened to determine Sindall's employment. Sindall drew attention to delays due to the issue of 123 instructions by the contract administrator and requested a further extension of time. They sent a box of files to the contract administrator and gave him seven days for a response. Michael Edwards asked for more time to consider the submission but the request was ignored and Sindall commenced adjudication. The court considered that, in view of the short period of time given to Michael Edwards to reach a decision, there could be no dispute which was referable to adjudication. The judge said:

'It must be clear that a point has emerged from the process of discussion, or the negotiations have ended, and that there is something which needs to be decided'.

12.3.3 Whether or not a dispute has in fact occurred will often depend upon the facts of the case. Remarks of judges in the following cases are very relevant:

(1) *Fastrack Contractors* v. *Morrison Construction* (2000) in which Judge Thornton QC stated:

'A dispute can only arise once the subject matter of the claim, issue or other matter has been brought to the attention of the opposing party and that party has had an opportunity of considering and admitting, modifying or rejecting the claim or assertion.'

(2) *Edmund Nuttall* v. *R.G. Carter* (2002) where Judge Seymour QC said:

'For there to be a dispute there must have been an opportunity for the protagonists each to consider the position adopted by the other and to formulate arguments of a reasoned kind.'

12.3.4   Other cases brought before the courts where the question of whether a dispute had arisen include:

Cowlin Construction Ltd v. *CFW Architects* (2002)
Costain Ltd v. *Westcol Steel Ltd* (2003)
Beck Peppiatt Ltd v. *Norwest Holst* (2003)
Orange EBS Ltd v. *ABB* (2003)

SUMMARY

For a dispute to have arisen which may be referred to adjudication it must be clear that the process of discussion or negotiation has ended and that there is something which needs to be decided.

12.4   To comply with the Construction Act and be subject to adjudication, the contract must be in writing or evidenced in writing. Would reference in meeting minutes to the nature of the work and the submission of fee accounts in relation to work undertaken by an engineer be regarded as a contract evidenced in writing?

12.4.1   Section 104 (2) of the Construction Act covers services provided by an engineer undertaking design work. The matter which had to be decided by the adjudicator in the case of *RJT Consulting Engineers Ltd* v. *DM Engineering (Northern Ireland) Ltd* (2002) was whether the contract complied with the requirement to be in writing or evidenced in writing.

The contract between RJT Consulting Engineers and DM Engineering was essentially oral. Both the parties to the dispute were involved with the refurbishment of the Holiday Inn in Liverpool. RJT's representative verbally agreed with a representative from DM to undertake some design work for a fee of £12 000. DM was the mechanical and electrical contractor and RJT the consulting engineer. A dispute arose whereby DM levied a claim for negligence in the sum of £858 000 against RJT. The matter was referred to adjudication by DM. RJT applied to the court for a declaration that the agreement was not in writing and therefore not covered by the Construction Act. DM argued that, whilst the contract was not in writing, it was evidenced in writing. RJT's case was that to be evidenced in writing the evidence must recite the terms of the agreement. Judge Mackay disagreed. He noted that the material concerning the contract was extensive. RJT had submitted fee accounts which identified the nature of the work, the names of the parties and the place of work. There were minutes of a meeting which again referred to parties and the nature of the work undertaken. The judge concluded that the evidence in support of the agreement would not be required to identify its terms and that due to the extensive number of documents the agreement came within the ambit of the Construction Act.

12.4.2 On appeal the Court of Appeal took a different view, as expressed by Lord Justice Ward when he made the following observations:

> 'On the point of construction of section 107, what has to be evidenced in writing is literally the agreement, which means all of it, not part of it. A record of the agreement also suggests a complete agreement, not a partial one.'

12.4.3 The Court of Appeal, in coming to this decision, seemed to be taking the view that there is no difference in meaning between a contract in writing and one evidenced in writing. This being the case, it was pointless a distinction being made in the wording of the Act.

12.4.4 The decision of the Court of Appeal was followed in *Debeck Ductwork Installations Ltd* v. *T. and E. Engineering Ltd* (2002) and *Tally Weigl (UK) Ltd* v. *Pegram Shopfitters Ltd* (2003).

SUMMARY

To comply with the Construction Act all of the agreement must be in writing. There seems to be no provision for a contract which is part in writing and part oral. It would be interesting to speculate what the attitude of the Court of Appeal would have been in the case of *RJT Consulting Engineers Ltd* v. *DM Engineering (Northern Ireland) Ltd* (2002) had those parts of the contract which were in dispute been in writing, with the remainder over which there was no dispute subject to an oral agreement.

12.5 Can a dispute concerning verbal amendments to a construction contract be referred to adjudication?

12.5.1 Section 107 of the Construction Act requires a contract to be in writing or evidenced in writing for it to be referable to adjudication. Where the contract itself is in writing does this provision apply to amendments to the contract?

12.5.2 A dispute concerning this matter arose in the case of *Carillion Construction* v. *Devonport Royal Dockyard* (2003). Carillion was contracted to undertake refurbishment at Devonport Royal Dockyard. The contract was in writing, under which Carillion was to be reimbursed its costs and a fee. A gain share/ pain share arrangement applied whereby any underspend compared with the target was shared by the parties and in like manner any overspend was shared. The target cost was, over the period of contract, increased from £56m to £100m. Carillion argued that an oral agreement had been reached to abandon the gain share/pain share with payment being made on a fully cost-reimbursable basis. An application for payment to accord with achieving milestone 33 was made by Carillion based upon its costs and a fee. The employer refused to make the payment and the dispute was referred to adjudication.

12.5.3   The adjudicator decided that a binding agreement had been entered into to amend the contract and that the employer was due to pay to Carillion the sum of £7 451 320 plus VAT. It was argued by the employer that there was no binding oral agreement and if one existed, as it had not been evidenced in writing, it fell outside the adjudicator's jurisdiction.

It was held by the court that the Construction Act does not provide for adjudication in respect of an oral variation to a written construction contract.

SUMMARY

The Construction Act does not provide for adjudication in respect of an oral variation to a written construction contract.

12.6   Where a mediator is appointed in relation to a dispute in connection with a construction contract and the dispute is not resolved but referred to adjudication, is the mediator barred from being appointed as adjudicator?

12.6.1   In the case of *Glencot Development & Design Co Ltd* v. *Ben Barratt & Son (Contractors) Ltd* (2001), the court was asked to decide whether there was evidence of bias on the part of the adjudicator which would prevent the enforcement of his decision. The claimant, Glencot, was a subcontractor to the defendant, Barratt, for the provision of 1200 mild steel wind posts which were provided as part of a brickwork subcontract. It was agreed by the parties that the value of the work was £390 000 plus VAT. The dispute, however, concerned whether Barratt was entitled to a 3% discount. Mr Peter Talbot was appointed as the adjudicator. In the first instance, at the request of the parties, he acted as a mediator. The mediation did not resolve the dispute and Mr Talbot reverted to his role of adjudicator. He wrote to both parties offering to withdraw from the adjudication if either party felt that his ability to make an impartial decision had been affected by his presence during the settlement negotiations. Barratt was of the opinion that Mr Talbot should withdraw. Mr Talbot took legal advice and decided to continue with the adjudication. The decision of Mr Talbot was that the final account should be £431 616 plus VAT, giving a balance due of £160 016 plus VAT, having taken into account amounts already paid.

12.6.2   The court had to consider whether to enforce the decision. It was argued that the decision should not be enforced due to bias on the part of the adjudicator. Bias can come in one of two forms. Actual bias is where it can be demonstrated that a judge, arbitrator or adjudicator is actually prejudiced in favour of or against one of the parties. Apparent bias may occur where circumstances exist which give rise to a reasonable apprehension that the judge, arbitrator or adjudicator may have been biased. The mediation process involved the mediator holding private discussions with each party without the other party being informed of what was said. It would appear that, in the process, heated discussions took place which may have given the impression that Mr Talbot had already formed a view. The court was also influenced by Mr Talbot's writing to each party offering to withdraw, which may lead to a conclusion that there was doubt in his mind. Under the circumstances the court refused to enforce the adjudicator's decision.

SUMMARY

Where a dispute is referred to a mediation process which fails to resolve the matter and the mediator is subsequently appointed as the adjudicator, it is unlikely that a court will enforce any decision the adjudicator may make. However, the decision of the court may depend upon the circumstances.

12.7 Will a court enforce an adjudicator's award which is clearly wrong?

12.7.1 Adjudicators are human and as such are likely to make errors. Where an error occurs in an adjudicator's decision, will the court refuse to enforce the decision, refer it back to the adjudicator for correction or order enforcement?

12.7.2 The case of *Bouygues UK Ltd* v. *Dahl & Jensen Ltd* (2000) addressed the question as to whether an adjudicator's award which contains an obvious error will be enforced by the court. A dispute arose in connection with claims for additional works, delay and disruption. There was a purported determination of the contract under an express term which led to a counterclaim.

The adjudicator found in part for both parties. In his decision he set out the original tender amount of £5 450 000 to which he added amounts for additional works and delay to give a total of £7 626 542. From this he deducted sums in respect of the counterclaim to arrive at a final sum of £6 979 912. This was a gross amount and included 5% retention. He then deducted the net amount paid which was £6 732 171, leaving a balance due of £247 741. It was agreed that, in deducting a net paid figure with no retention from a gross figure which included retention, the adjudicator had made an error.

No payment was made and enforcement proceedings were commenced. The judge in the case said he had to decide how a mistake should be characterised. In providing a decision regarding a matter which was not referred to him, the adjudicator acts outside his jurisdiction and any decision would not be enforced. If, however, he decides on a matter which is referred to him but his decision is mistaken, then the decision would be enforced. In other words, if the adjudicator is doing what he is asked to do and is answering the right questions but giving the wrong answer, his decision would be enforced by the court. In this case the adjudicator answered the right question but gave the wrong answer and therefore it was enforced by the court.

12.7.3 The case of *Shimizu Europe Ltd* v. *Automajor Ltd* (2001) also dealt with an adjudicator's error. Automajor contracted with Shimizu for the design and construction of an office under a JCT 98 With Contractor's Design contract. A dispute arose concerning set-off, extensions of time, the status of the agreement and release of retention. A further matter in dispute related to variations to the smoke ventilation works. The adjudicator ruled that there had been no variation to the smoke ventilation works but nonetheless included a sum of £161 996 in his decision in respect of this matter. This was an obvious error but nonetheless the decision was enforced by the court.

SUMMARY

Whether a court will enforce an adjudicator's award which contains an error will depend upon the type of error involved. If a matter is referred to an adjudicator who, in his decision, arrives at the wrong answer, the decision will be enforced. Should the adjudicator's decision include a matter which was not referred to him it will be regarded as being outside his jurisdiction and will not be enforced.

12.8 If a dispute is the subject of ongoing litigation, can one of the parties, whilst the litigation is in progress, refer the matter to adjudication?

12.8.1 Litigation can be a slow process, particularly when it involves complex construction matters. It is understandable therefore if one of the parties, usually the claimant, becomes impatient and considers commencing adjudication proceedings. Would a court be prepared to intervene to stop the adjudication? If both judge and adjudicator were to make a decision with regard to the same matter it is a distinct possibility that their decisions would, in some marked way, differ.

12.8.2 In the case of *Herschel Engineering Ltd* v. *Breen Properties* (2000), county court proceedings were commenced for the recovery of sums alleged to be due under a construction contract. Despite the existence of court proceedings, the dispute was also referred to adjudication. It was the defendant's opinion that adjudication was inappropriate as the matter was already the subject of litigation.

The adjudicator made an award in the claimant's favour and as the defendant refused to pay, the matter was referred to court. It is well established law that the same dispute cannot be heard before two legal forums such as a court and arbitration. The court, however, decided that this rule does not apply to adjudication under the Construction Act.

12.8.3 The judge explained that Parliament's intention in enacting the statute was to produce a speedy mechanism for resolving disputes arising out of construction contracts on a provisional and interim basis. Decisions of adjudicators are to be enforced pending final determination by arbitration, litigation or agreement. He considered that, if Parliament had intended that a party should not be able to refer a dispute to adjudication once litigation or arbitration had commenced, it would have been expressly stated.

The claimant was therefore successful in recovering the amount of the adjudicator's decision. If the defendant was successful in the county court action, the claimant would have to repay the sum awarded by the adjudicator. Had the position been otherwise the judge said he might not have enforced the adjudicator's award.

SUMMARY

In the case of *Herschel Engineering Ltd* v. *Breen Properties* (2000) it was held that there was nothing to prevent litigation and adjudication from proceeding on the same matter at the same time. The judge in the case was of the opinion that, if Parliament had intended that a party should not be able to refer a dispute to adjudication once litigation or arbitration had commenced, it would have been expressly stated.

12.9 Can an adjudicator withhold a decision from the parties until his fees are paid?

12.9.1 Most adjudicators operate on a commercial basis and require payment for their services. It is common practice in the commercial world for a supplier to require payment before delivery, which helps to ensure that, at the end of the financial year, write-off for bad debt is kept to a minimum. Adjudicators know the uncertainties attached to receiving remuneration for their efforts and some insist upon payment of their fees before releasing a decision.

12.9.2 In the case of *St Andrew's Bay Development* v. *HBG Management and Mrs Janey Milligan* (2003), a dispute concerning the construction of a leisure centre was referred to adjudication. To ensure that the statutory time-scale of 28 days for reaching a decision was adhered to, the adjudicator's decision was required not later than 5 March 2003. Sometime after 5 PM on that day, HBG contacted the adjudicator to enquire when a decision could be expected. They were advised that the adjudicator had reached a decision but it would not be released until her fee had been paid. A fee account was issued later that day and on the following day HBG undertook to pay the whole of the fee.

12.9.3 The adjudicator conveyed her decision by fax on 7 March 2003, which was two days late. The court had to decide whether the decision, due to it being two days late, was invalid and whether the adjudicator is entitled to withhold the release of a decision until the fees have been paid. The statute makes no reference to communicating the decision but the conditions of contract required the decision to be issued 'forthwith'. This, explained the judge, meant immediately. However, it was held that whilst the two-day delay was serious, it was not sufficient to render the decision a nullity.

12.9.4 With regard to the payment of fees, the judge ruled that an adjudicator is not entitled to withhold the decision until the fees have been paid. This difficulty could be overcome if the parties entered into an agreement whereby the adjudicator could withhold fees. Most adjudicators now include such a clause in their terms of appointment. Even when agreement is reached on this matter, the adjudicator is not entitled to withhold the issue of the decision beyond the 28 days statutory period, whether or not the fees have been paid.

SUMMARY

An adjudicator can withhold the issue of a decision until the fees have been paid only if the parties agree. Most adjudicators now include such a term in their conditions of appointment. However, even when agreement has been reached, the adjudicator cannot withhold the decision beyond the statutory period.

12.10 If an adjudicator issues a decision late, can it be enforced?

12.10.1 The Scheme for Construction Contracts provides for an adjudicator to make a decision within 28 days of the referral notice. In the case of *Ritchie Brothers (PWC) Ltd* v. *David Phillips (Commercials)* (2004) it was argued that, as the adjudicator was late in making a decision, it was unenforceable.

12.10.2 The court had to decide in the first instance the exact date from which the 28-day period began to run. A referral notice dated 18 September 2003 was posted on that day to the adjudicator. The letter could not be delivered on 19 September 2003 as it was a public holiday in Edinburgh, but was collected from the postal sorting office by the adjudicator on 23 September 2003. On 23 October 2003 the adjudicator wrote to the parties requesting payment of his fees. The pursuers agreed and the adjudicator delivered his decision on 27 October 2003. It was contended by the defendant that the 28-day period began to run on 18 September 2003 and expired on 16 October 2003. The defendant's case was that the period began to run on 23 September 2003, the date the adjudicator collected the referral notice from the postal sorting office. Even if the defendant was correct, the decision delivered on 27 October 2003 was late.

12.10.3 The court took the view that the adjudication process does not end 28 days after the referral. Provision is made in the Scheme for the adjudicator to be replaced if the one originally appointed does not produce a decision. This being the case, a late decision would still be enforceable provided another adjudicator had not been appointed. It was the view of the court that the 28-day period is directory only and not mandatory.

12.10.4 A different approach to the matter was taken in the earlier case of *Barnes & Elliott Ltd* v. *Taylor Woodrow Holdings and George Wimpey Southern Ltd* (2003). The court in this case was influenced by the decision in *St Andrew's Bay Development Ltd* v. *HBG Management Ltd and Mrs Janey Milligan* (2003), which drew attention to the provisions of the Scheme for Construction Contracts under which the decision is to be made within 28 days with no reference as to the time for its communication. In *Barnes & Elliott* the decision was communicated two days outside the 28-day period. The court enforced the adjudicator's decision which was considered to have been made within the 28-day period, although delivered two days late.

12.10.5 The court, in *Simons Construction Ltd* v. *Aardvark Developments Ltd* (2003), followed a similar line to that in *Ritchie Brothers (PWC) Ltd* v. *David Phillips (Commercials) Ltd* (2004). It was held that a late adjudicator's decision was not invalid unless a fresh notice of referral had been issued before the release of the adjudicator's decision.

SUMMARY

Following the decision in the case of *Ritchie Brothers (PWC) Ltd* v. *David Phillips (Commercials) Ltd* (2004), a court will enforce a late adjudicator's decision provided a replacement adjudicator has not been appointed.

12.11 The Construction Act states that 'a party to a construction contract has the right to refer a dispute arising under the contract for adjudication'. As this suggests disputes can only be referred one at a time, does this mean that a dispute regarding the variations and also delays will have to be the subject of separate references?

In the case of *Fastrack Contractors Ltd* v. *Morrison Construction Ltd* (2000), Judge Thornton held that the dispute is whatever claims, heads of claim, issues, contentions or causes of action are in dispute and included in the adjudication reference. Lord MacFadyen, however, sounded a word of caution that it may be necessary for the adjudicator to consider whether there is sufficient connection between the disputed matters for them to be dealt with as a single dispute.

SUMMARY

Judge Thornton held that claims, heads of claim, issues, contentions or causes of action which are in dispute and included in a reference to the adjudicator will all be classified as a dispute.

12.12 Where a compromise agreement relating to a dispute on a construction contract is itself the subject of a dispute, can it be referred to adjudication?

In *Lathom Construction Ltd* v. *AB Air Conditioning* (2000), a dispute arose between the employer and contractor in relation to the construction of a factory in Skelmersdale. The matter was referred to an adjudicator appointed by the RICS. Before the adjudicator commenced work the parties reached a compromise agreement which was committed to writing. Unfortunately a dispute arose out of the compromise agreement due to a disagreement as to its meaning. The contractor requested the RICS to appoint an adjudicator and the original one was reappointed.

It was argued by AB Air Conditioning, the employer, that the compromise agreement was not a construction contract and therefore not catered for by the Construction Act. The adjudicator disagreed and proceeded to make a decision. The employer refused to comply with the adjudicator's decision and the contractor referred the matter to the court for enforcement. The matter before the court was whether or not the adjudicator had jurisdiction to act. The court refused to enforce the adjudicator's decision as it was considered the adjudicator had no jurisdiction to deal with a compromise agreement.

SUMMARY

A compromise agreement which arises out of a dispute on a construction contract cannot be referred to adjudication.

12.13 A matter in dispute can only be referred to adjudication once. Where an adjudicator's decision has been received relating to a dispute over sums included in an interim certificate in respect of variations, can a dispute relating to the value of those variations when included in the final account be referred to adjudication?

12.13.1 The Scheme for Construction Contracts applies in respect of an adjudication under the Act where the terms of the contract are not in compliance with the Act. Paragraph 9(2) of the Scheme stipulates that an adjudicator must resign where the dispute referred to him is substantially the same as one that has previously been decided by another adjudicator.

12.13.2 A dispute concerning this paragraph was referred to the court in the case of *Sherwood & Casson Ltd* v. *MacKenzie* (1999). Sherwood was a cladding contractor for the construction of a grandstand at Barrow RUFC. MacKenzie was the main contractor. The non-payment for variations in interim payments became a dispute and was referred to an adjudicator who awarded £6550 plus VAT to Sherwood.

The claim was repeated in the final account backed up by substantially more supporting information. Sherwood also submitted a claim for loss and expense. MacKenzie argued that the matter had already been referred to adjudication and therefore under paragraph 9(2) of the Scheme for Construction Contracts, the second adjudicator should resign. The adjudicator disagreed and rejected the loss and expense claim, but allowed a further £12 000 in respect of variations. MacKenzie refused to pay and the matter was referred to the court. It was held that the adjudicator can rule on his own jurisdiction but his decision could be raised in the enforcement proceedings.

Judge Thornton considered that, when reviewing the question of the adjudicator's jurisdiction, the court should give considerable weight to the adjudicator's decision on the matter. In finding that the two disputes were not substantially the same, he was influenced by the loss and expense claim and the information supporting the variation claim. The judge in passing said that his decision would have been the same if the adjudication rules issued by one of the institutions applied.

SUMMARY

A matter which has been referred to adjudication once cannot be referred a second time. However if, as in the case of *Sherwood & Casson Ltd* v. *MacKenzie* (1999), the matter in dispute has been the subject of amendment, it may be referred to adjudication a second time. Whether or not a second adjudication is permitted will depend upon the facts of the case.

12.14 Will a clause in a construction contract which states that all disputes are subject to the exclusive jurisdiction of the Austrian Courts and Austrian Law, or the courts and law of another jurisdiction outside the UK, result in disputes falling outside the UK courts' power to enforce an adjudicator's award?

12.14.1 There have been many attempts to circumvent the effects of the adjudication process under the Construction Act. In the case of *Comsite Projects Ltd* v. *Andritz AG* (2003), a clause in the contract stated that all disputes are subject to the exclusive jurisdiction of the Austrian courts and Austrian law.

12.14.2 A dispute arose between the parties which was referred to adjudication. Enforcement proceedings were commenced in respect of the adjudicator's decision. Despite the wording of the clause, the court ordered that the parties must comply with the adjudicator's decision. However, having complied with the adjudicator's decision, either party could refer the dispute to the Austrian court for a final decision.

SUMMARY

Attempts have been made to circumvent the effects of the adjudication process under the Construction Act. A clause written into the contract which states that all disputes will be subject to the exclusive jurisdiction of the Austrian courts and Austrian law, or the courts and law of any other jurisdiction outside the UK, will not affect the enforcement of an adjudicator's decision by the UK courts.

12.15 Are the parties entitled to challenge an adjudicator's fees on the grounds that they are unreasonably high?

12.15.1 Most adjudicators ensure that the basis on which their fees are to be calculated are agreed by the parties as soon as possible. It is usual for the adjudicator to charge fees on the basis of an hourly rate or a day rate. Whilst the rates may be agreed, the parties may consider that the number of hours or days included in the adjudicator's fee account is excessive.

12.15.2 The case of *Stubbs Richards Architects* v. *WH Tolley & Son Ltd* (2001) is an example of the parties challenging an adjudicator's fees.

Tolley & Son carried out construction work for Torridge District Council and as a result disputes arose. The matter was referred to adjudication and a partner of Stubbs Richards, a firm of architects, was appointed as the adjudicator. Two disputes were referred to the adjudicator who delivered his decision along with his fee account in the sum of £1500 plus VAT. The fee was paid but Tolley & Son alleged that it was excessive in that the number of hours which the adjudicator claimed to have worked could not be justified. An action was commenced to recover some of the fees which Tolley & Son had paid.

The district judge who heard the case considered that an adjudicator does not have immunity from a claim that the hours and remuneration are unreasonable. He ordered that some of the fees should be repaid to Tolley & Son. This decision was the subject of an appeal. On appeal a different view was expressed in finding for the adjudicator. The court drew attention to the wording of the Construction Act where it is stated that the adjudicator is not liable for anything done or omitted in the discharge of his or her functions unless carried out in bad faith. It was considered that the adjudicator's fee formed an integral part of his agreement and was therefore caught by the wording in the Act. The district judge was criticised for basing his judgment on what he considered would be the amount of time likely to have been spent by the solicitor in presenting a case before a court. This was felt to be inappropriate as an adjudicator would spend a lot more time reading through files, interviewing the parties and visiting the site than would a solicitor in presenting a case in court.

SUMMARY

The parties to an adjudication cannot challenge an adjudicator's fees. Under the Construction Act the adjudicator is not liable for anything done or omitted in discharge of his duties unless carried out in bad faith.

12.16 Does a draft adjudicator's decision constitute a final decision?

12.16.1 There seems to be a growing trend among some adjudicators to issue a draft decision to the parties and invite comments. This may show a lack of confidence on the part of those adjudicators who indulge in this practice or perhaps they like to err on the side of caution.

12.16.2 In the case of *Simons Construction Ltd* v. *Aardvark Developments Ltd* (2003), Simons were engaged as the main contractor under a JCT main contract. A dispute arose which was referred to adjudication. The 28-day period within which the adjudicator is required to produce an award was extended by agreement of the parties to a decision required by 17 June 2002. A draft decision was issued on that day inviting the parties to offer their comments. A final decision was given a week later. It was argued by Aardvark, the successful party, that the draft decision was a valid decision under the Act. Simons argued that the draft decision was not a final decision and as the final decision was given late it was invalid. The court held that the draft decision was not intended to be a final decision and as such did not comply with the Act. However, as no fresh notice of referral had been issued by either party, the final decision stood as valid even though it was issued late.

SUMMARY

Where an adjudicator issues a draft award, the court will look at his intentions when deciding whether or not it can be considered to be a final award. It would seem unlikely, except in the most unusual of cases, for an adjudicator to issue a draft award which he considers to be a final award.

12.17 Does an adjudicator who communicates with one party without disclosing the details of the communication to the other party risk his decision being nullified by the courts on the grounds that his conduct amounted to bias?

12.17.1 The rules of natural justice provide that all persons who are accused are entitled to know the charges made against them and must be given a proper opportunity to be heard. These rules have been examined and re-examined, interpreted and misinterpreted by judges and legal commentators down the years. Arbitrators have been almost paranoid in ensuring that these rules are not contravened, by giving parties and their advisers every opportunity to be heard and seeing that arbitrators do not hear evidence from one party without giving the other party a proper opportunity to respond. Are the decisions of adjudicators also vulnerable from a failure to comply with the rules of natural justice?

12.17.2 In the case of *Discain Project Services* v. *Opecprime Developments* (2000), the court had to decide whether an adjudicator had contravened the natural justice rules. A payment dispute arose which was referred to adjudication. The adjudicator's decision was that the respondent should pay the applicant a sum of £55 552. No payment was made, on the grounds that there had been a breach of the rules of natural justice, and the matter was referred to court. It appears that the adjudicator had a private telephone conversation concerning the adjudicator's jurisdiction, the validity of the Application for Payment and the withholding notice. This was followed by a fax from the adjudicator which stated his intention to proceed to a decision without the benefit of further comment from the parties. A second telephone conversation between the adjudicator and applicant took place after the fax had been sent.

The court refused to enforce the adjudicator's decision. In the opinion of the judge there was a very serious risk of bias and there were clear failures to consult with one of the parties on important submissions made by the other party. The adjudicator should also have made sure he involved both parties in discussions concerning his jurisdiction, which he failed to do.

12.17.3 Adjudicators, under the rules included in The Scheme for Construction Contracts, are allowed to make their own enquiries to ascertain the facts concerning the matters in dispute. There is, however, a fine line to be drawn between making those enquiries and breaching the rules of natural justice.

SUMMARY

Adjudicators are allowed to make their own enquiries to ascertain the facts relating to the matters in dispute. However, if ascertaining the facts involves communicating with one of the parties, an adjudicator must keep the other party fully informed.

12.18 Can an adjudicator employ the services of an expert to assist in making a decision?

12.18.1 Adjudicators, in their terms of appointment, are usually given power to engage the services of an expert in arriving at a decision. When taking a decision to appoint an expert, the adjudicator's path does not always run smoothly.

12.18.2 In the case of *Try Construction Ltd* v. *Eton Town House Group Ltd* (2003), an adjudication was commenced following which a meeting took place between the parties and the adjudicator. At this meeting the parties agreed that the adjudicator could appoint a programming expert, a Mr Lowsley. During a second meeting both parties agreed that Mr Lowsley could contact their respective programming experts independently.

12.18.3 The adjudicator, in his decision, awarded an extension of time to Try and the payment of £269 916 plus interest. Eton refused to pay on the grounds that the adjudicator, in using the information provided by Mr Lowsley, had adopted a methodology which they had not had a chance to consider and comment upon. Try argued that Eton had agreed to the appointment of the expert and they could not therefore raise an objection after the adjudicator had reached a decision.

12.18.4 The court, in upholding the adjudicator's decision, held that the evidence showed that the defendant had participated in the adjudication process without demur and knew that Mr Lowsley, as an expert programmer, would investigate the claim on his terms. The defendant had not requested a preliminary or draft finding as the basis for further submissions. The court accepted evidence that it had been agreed at the parties' first meeting with the adjudicator that Mr Lowsley would use his own expertise in the delay analysis.

12.18.5 A different outcome occurred in the case of *RSL (South West) Ltd* v. *Stansell Ltd* (2003). This case involved a dispute between a contractor and sub-contractor relating to the fabrication and erection of structural steelwork and staircases for a project in Union Street, Bristol. The dispute included delay issues and the adjudicator required someone with planning experience to assist him with the programming matters. Both parties agreed to the adjudicator seeking assistance and Stansell requested sight of any report which might be prepared by the expert, and reasonable time to pass comment.

The expert prepared a report which was passed to the parties for comment. RSL made a series of comments and submitted further information as they did not find the report very favourable. Stansell, obviously pleased with the report, passed no comments. Following receipt of RSL's comments the expert produced a final report. Unfortunately the adjudicator did not show the final report to the parties and RSL therefore contested the decision.

12.18.6 Judge Richard Seymour QC considered the complaint by RSL justified. The adjudicator should have disclosed the final report to the parties and in failing to do so was in breach of the rules of natural justice.

12.18.7 In *Balfour Beatty Construction Ltd* v. *The Mayor and Burgesses of the London Borough of Lambeth* (2002), the court held that the adjudicator had failed to comply with the rules of natural justice. It seems the adjudicator was not impressed with the details concerning an extension of time entitlement submitted by Balfour Beatty Construction Ltd. He decided to employ the help of two of his colleagues to construct a collapsed as-built analysis to assist him in arriving at a decision. This information was not passed to the parties for comment and as a result the court refused to enforce the adjudicator's decision on the grounds of bias and breach of the rules of natural justice.

SUMMARY

If an adjudicator wishes to use the services of an expert he must seek the agreement of the parties. In addition, any report produced by the expert must be passed to both parties to allow them to pass comment.

12.19 Where an adjudicator seeks legal advice to assist him in reaching a decision, is he obliged to reveal the advice to the parties involved in the adjudication?

12.19.1 Many adjudicators are neither legally qualified or trained. Despite this lack of knowledge and training, adjudicators are regularly required to deal with legal matters as part of their duties. It is not uncommon, therefore, for adjudicators to seek independent legal advice to assist them in reaching a decision. Most adjudicators will include a term in their conditions of appointment which gives them authority to acquire legal opinions or there may be provision within the contract between the parties which deals with the matter.

12.19.2 This was the situation in the case of *BAL* v. *Taylor Woodrow Construction Ltd* (2004). A dispute arose out of a contract Taylor Woodrow undertook on behalf of Nationwide Anglia Property Services Ltd and involved BAL one of the subcontractors. Nationwide issued a letter of intent giving Taylor Woodrow an instruction to commence work. The letter of intent placed a cap of £750 000 on the value of work to be carried out without any further instructions. The subcontract included a clause which allowed any adjudicator who may be appointed to obtain advice from specialists provided one of the parties requests or consents and agrees to be responsible for the costs and expenses incurred. A dispute arose between Taylor Woodrow and BAL, which was referred to adjudication. The adjudicator indicated to the parties that he may require to seek legal advice concerning the effect of the cap in the letter of intent. BAL agreed and Taylor Woodrow did not object. The adjudicator did not inform the parties as to what information he provided to his legal adviser or the advice he was given.

Taylor Woodrow argued that there had been a breach of the rules of natural justice as the adjudicator had not disclosed to the parties the legal advice he had received. The court agreed and refused to enforce the adjudicator's decision.

SUMMARY

Where an adjudicator receives legal advice to enable him to reach a decision, he should disclose it to the parties. If he fails to do so the courts are likely to consider this failure to be a breach of the rules of natural justice, and to refuse to enforce the adjudicator's decision.

12.20 Have the courts laid down any general guidelines as to how the rules of natural justice should be applied in respect of adjudication?

12.20.1 The rules of natural justice have been applied over many years to ensure the parties get a fair hearing. Often courts appear to bend over backwards to see that fairness applies. Basically, the rules of natural justice are aimed at ensuring that both parties know fully the case which is being made against them and are given a proper opportunity of providing answers. All courts of law, arbitrations, tribunals and adjudication are required to apply the rules of natural justice.

12.20.2 In the case of *Costain Ltd* v. *Strathclyde Builders Ltd* (2004), Strathclyde Builders refused to comply with an adjudicator's decision on the grounds that during the procedure there had been a breach of the rules of natural justice.

The adjudicator arrived at a decision following discussions with his legal adviser. However, he failed to disclose to the parties the details received from the lawyer. Like the decision in *BAL* v. *Taylor Woodrow Construction Ltd* (2004), it was held by the court that the actions of the adjudicator amounted to a breach of the rules of natural justice.

12.20.3 The court then took the opportunity to formulate guidelines which should be applied by adjudicators to ensure they do not fall foul of the rules of natural justice:

(1)   Each party must be given a fair opportunity to present its case. That is the overriding principle, and everything else is subservient to it.

(2)   Subject to (1) above, together with any express provisions in the parties' contract, procedure is entirely under the control of the adjudicator.

(3)   In considering what is fair, it must be remembered that adjudications are conducted according to strict time limits; thus time limits will be severely restricted.

(4)   The adjudication procedure is designed to be simple and informal; all that it will normally require is that each party should be given an opportunity to make comments at any relevant stage of the adjudication process.

(5)   Each party must be given an opportunity to comment on the other's submission and additional new submissions.

(6)   An adjudicator is normally given power to use his own knowledge and experience in deciding the issues in dispute. If he raises new points that have not been canvassed by the parties, it will normally be appropriate to invite submissions.

(7)   If the adjudicator obtains more information or carries out tests, it will be appropriate to make the results known to the parties, and call for their comments.

(8)   An adjudicator may be given power to obtain from other persons such information and advice as he considers necessary on technical or legal matters. If he does so the situation is similar to (6) above.

(9)   In this connection, no distinction can be drawn between issues of fact and issues of law.

SUMMARY

The rules of natural justice apply to adjudication. To ensure compliance the adjudicator must ensure that both parties know fully the case which is being made against them and are given a proper opportunity of providing answers. In the case of *Costain Ltd* v. *Strathclyde Builders Ltd* (2004), the court held that, in failing to disclose to the parties legal advice he had received, the adjudicator had breached the rules of natural justice. In arriving at a decision the court took the opportunity of producing guidelines on how the adjudication process should be conducted to ensure compliance with the rules of natural justice.

12.21 Where an adjudicator issues a decision involving the payment of a sum of money by the winning party to the losing party, can the amount paid be reduced or eliminated in compliance with a clause in the contract?

12.21.1 When coming to a decision as to whether a set-off can be made from an adjudicator's award, the courts have to decide whether the enforcement of an adjudicator's decision is based upon a separate statutory force or no more than a contractual obligation which is backed by statute. When the question as to whether an adjudicator's award could be reduced or eliminated as a result of contractual entitlement came before the courts, the view taken was that an adjudicator's decision creates a contractual obligation which is backed by statute.

12.21.2 In *Parsons Plastics* v. *Purac Ltd* (2002), a contract between a main contractor and subcontractor contained a clause which expressly preserved the contractor's common law rights of set-off. The Court of Appeal held that an adjudicator's decision was subservient to this clause, allowing the contractor to set off entitlements against the sum included for payment to the subcontractor in the adjudicator's decision.

12.21.3 This line of argument was followed in the case of *Bovis Lendlease* v. *Triangle Developments* (2003), which dealt with the payment provisions where a contractor's employment is terminated. Clauses which deal with this situation in most standard forms of contract state that following termination no further payment needs to be made until the final account and claims have been resolved. In this case an adjudicator's decision was not immediately honoured but made subject to the provisions of the termination clause.

12.21.4 The Court of Appeal made an about-turn in the case of *Levolux* v. *Ferson* (2003). In a similar fashion to the *Bovis Lendlease* case, the decision involved a termination but in this case it was a subcontractor whose employment was terminated. A dispute arose concerning the subcontractor's right to payment. The dispute was referred to adjudication and a decision made in favour of the subcontractor. Due to the fact that the subcontractor's employment had been terminated, the contractor considered that, following the provisions of the termination clause in the contract, he was under no obligation to make payment. The Court of Appeal disagreed in holding that to allow the contractor to withhold payment of the adjudicator's decision would defeat the intended purpose of adjudication under the Construction Act. The contract must be interpreted to achieve this purpose and any offending clause would be struck down.

12.21.5 A slight twist to the situation occurred in *Shimizu Europe Ltd* v. *LBJ Fabrications Ltd* (2003). In this case a dispute arose in connection with cladding work at an Oxford Science Park. The adjudicator, in his decision, provided for Shimizu to pay LBJ the sum of £47 000 plus VAT 'without set off'. Payment was to be made by Shimizu not later than 28 days after LBJ had delivered a VAT invoice.

After the issue of the VAT invoice but before the payment was required to be made, Shimizu issued a withholding notice for set-off in respect of alleged defective work. The court allowed Shimizu to withhold the money on the grounds that the withholding notice complied with the contract and the money was not due for payment until 28 days after the issue of the VAT invoices.

SUMMARY

The Court of Appeal has changed its position in respect of whether there is a right of set-off from an adjudicator's decision in compliance with a clause in the contract. In *Parsons Plastics* v. *Purac Ltd* (2002), the Court of Appeal held that an adjudicator's decision is subservient to a clause under the contract which allows for set-off. A shift of position occurred in the decision of *Livolux* v. *Ferson* (2003) when the Court of Appeal stated that to allow a set-off from an adjudicator's decision would defeat the intended purpose of adjudication under the Construction Act.

12.22 Where an adjudicator's decision provides for a sum of money to be paid by the employer to the contractor, can the employer deduct from the sum awarded liquidated and ascertained damages due under the contract?

12.22.1 The case of *The Construction Centre Group* v. *The Highland Council* (2003) dealt with an adjudicator's decision arising out of a design, construction and maintenance contract on the west coast of Scotland. Disputes arose which were referred to an adjudicator whose decision provided for the employer to pay the contractor a sum in the region of £250 000. No payment was made as the employer claimed that the contractor owed liquidated damages in the sum of £420 000.

12.22.2 Within 7 days of the adjudicator's decision, the employer served a with-holding notice under section III of the Construction Act in respect of the liquidated damages. There had been no earlier set-off notice as the previous valuation was nil and therefore no sum of money was due from which the liquidated damages could be deducted.

The Outer House of the Court of Session in Scotland did not accept the employer's arguments. It was considered that the employer could have raised the matter of liquidated and ascertained damages in its submission to the adjudicator, which would provide an opportunity of having them dealt with in the decision. The employer was not at liberty to make a deduction from the adjudicator's award.

12.22.3 In *A* v. *A* (2002) the adjudicator awarded an extension of time for 46 weeks out of 112 claimed. The adjudicator was implicitly ruling that there was a delay of 66 weeks to completion, which would attract liquidated and ascertained damages of £75 000 per week. The employer sought to set off the liquidated damages from the sum awarded, but the court disagreed. He was obviously in error in not including a claim for liquidated and ascertained damages as part of the submission to the adjudicator, and therefore the court enforced the adjudicator's decision in full.

SUMMARY

An employer cannot deduct liquidated damages due under a contract from an adjudicator's award. If the employer considers he has an entitlement to liquidated damages, reference to the entitlement should be included in the employer's submission to the adjudicator.

12.23 Can a party to a construction contract, who is reluctant to have a dispute referred to adjudication, successfully argue that the adjudicator's decision should not be enforced as, due to the very restricted nature of the adjudication process, it is contrary to the European Convention on Human Rights?

12.23.1 There has been a great deal of speculation that Article 6 of the European Convention on Human Rights which came into force in the UK on 1 October 2000 has the effect of outlawing statutory adjudication. Article 6 states that each party must be afforded a reasonable opportunity to present his case, including his evidence, under conditions that do not place him at a substantial disadvantage vis-à-vis his opponent. An action was commenced in the case of *Elanay Contracts Ltd* v. *The Vestry* (2000) to enforce an adjudicator's award.

The defendant's case was that enforcement should not be ordered as the adjudication procedure was unfair. This was partly due to the fact that for much of the time the defendant's key witness was attending hospital visiting his dying mother. The shortness of the proceedings, which were over in 35 days, added to the sense of unfairness. Judge Richard Havery took a different view. He considered that Article 6 does not apply to statutory adjudication. The reason he gave was that, whilst proceedings before an adjudicator determine questions of civil rights, they do not provide for a final determination. An adjudicator's decision is provisional in that the matter can be reopened for a final decision in a subsequent court hearing or before an arbitrator.

12.23.2 In the case of *Austin Hall Building* v. *Buckland Securities* (2001), the losing defendant refused to comply with the decision of the adjudicator on the grounds that there had been a breach of the Human Rights Act. The basis of the case was that the adjudication process under the Construction Act was unfair and therefore contravened the European Convention on Human Rights. The defendant's line of argument was in respect of the 28-day period within which the adjudicator made his decision.

This did not get very far with the judge. Whilst the Human Rights Act states that it is unlawful for a tribunal to act in a way which is incompatible with a Convention right, it goes on to add that this requirement does not apply if, under a provision of primary legislation, the tribunal could not have acted differently. The adjudicator was obliged under the Construction Act to give a decision within the 28-day period and therefore it fell within the exception to the requirement of the Convention right. The judge went on to consider whether an adjudicator is in fact a tribunal. He felt that, as an adjudicator is not involved in legal proceedings, he is not therefore a tribunal.

SUMMARY

It is clear from the decisions in *Elanay Contracts Ltd* v. *The Vestry* (2000) and *Austin Hall Building* v. *Buckland Securities* (2001) that the adjudication process does not contravene the European Convention on Human Rights.

12.24 Can the party that is successful in adjudication recover its costs from the losing party?

12.24.1 Where disputes are referred to litigation or arbitration, the judge or arbitrator has power to include in the award that the losing party is to pay the winning party's costs. Experience has shown that the costs incurred by a party in referring a dispute to adjudication are significantly less than would be the case if the matter were referred to arbitration or litigation. Nonetheless the costs incurred can be substantial.

12.24.2 A party whose dispute when referred to adjudication is successful will no doubt be looking to the losing party for reimbursement of the costs incurred. Does an adjudicator have the same powers as a judge or arbitrator when it comes to awarding costs?

12.24.3 In the case of *Northern Developments (Cumbria) Ltd* v. *J. & J. Nichol* (2000), Northern Developments Ltd was the main contractor for the construction of an outlet for M. Sport Ltd at Doventry Hall in Cumbria. J. and J. Nichol was a subcontractor appointed to construct the steel frame, roofing, cladding and associated works. The terms of the subcontract were DOM/2 and the contract was governed by the Construction Act 1996. As the terms of the subcontract did not comply with the provisions of the Act, the Scheme for Construction Contracts applied.

On 13 July 1999 the subcontractor made an application for payment and a set-off notice was issued by the main contractor on the same day. The notice referred to defective work and delays. No payment was made to the subcontractor as the contractor argued that the set-off for defective work exceeded the value of the work executed. The subcontractor withdrew from site on 6 August 2000. This was treated by the contractor as a repudiatory breach which it accepted and appointed another contractor to finish off the work.

The matter was referred to adjudication by the subcontractor, who claimed an entitlement to a sum of £237 120 plus VAT together with £11 456 repayment of cash discount. The main contractor claimed set-off with regard to:

(1) Defective work
(2) Delays
(3) Damages due to the subcontractor's repudiation of the contract.

No reference was made in the set-off notice to the damages relating to repudiation as the withdrawal from site by the subcontractor took place after the issue of the notice.

The adjudicator had to decide in the first instance whether he was empowered to deal with the matter of repudiation. His decision was that he had no power to deal with the repudiation issue, but was limited to deciding matters which were under the contract. The repudiation issue did not, in his opinion, constitute a matter under the contract.

The judge considered that the adjudicator was wrong in not dealing with the repudiation. Acceptance of repudiation brings to an end the performance of the contract. The contract still exists and rights arising under it can be enforced. Nonetheless, the damages associated with the repudiation could not be set off as they had not been referred to in the notice. The adjudicator's decision was correct but for the wrong reasons.

12.24.4 Costs were awarded by the adjudicator in favour of the subcontractor but this was challenged by the contractor. In the case of *John Cothliff Ltd* v. *Allen Build (North West) Ltd* (1999), it was held in Liverpool County Court that an adjudicator had power to award costs under the Scheme Rules. The judge in the *Northern Developments* case disagreed with this reasoning as he considered that the Scheme gave the adjudicator power to deal with case management matters only. If Parliament had intended to give power to award costs, it would have said so. However, the judge considered that the parties could agree to give the adjudicator the power to award costs. In the *Northern Developments* case the parties had in fact given the adjudicator the power to award costs, as both of them had asked for costs in the referral document and response.

SUMMARY

An adjudicator has no power to provide in his decision that the losing party is to pay the winning party's costs. However, both parties may agree to vest the adjudicator with such powers.

In the case of *Northern Developments (Cumbria) Ltd* v. *J. & J. Nichol* (2000), it was held that, as both parties included in their submission that the opposing party should pay its costs, this in itself gave the adjudicator the necessary powers.

12.25 Where a contractor includes in its subcontract terms a clause which states that if a dispute is referred to adjudication the subcontractor, even if successful, will be liable to pay the contractor's costs, is such a clause enforceable?

12.25.1 It would seem very unfair for a main contractor to include a clause in its subcontract conditions to the effect that, if a dispute between the parties is ever referred to adjudication, the subcontractor will always be obliged to pay the costs of employing the adjudicator and the main contractor's own costs.

The provision is expressed as applying even if the adjudicator's decision goes in favour of the subcontractor. Would a court enforce such a clause?

12.25.2 In the case of *Bridgeway Construction* v. *Tolent Construction* (2000), the contractor included in the subcontract conditions a clause which required the subcontractor to pay his own and the main contractor's costs even if the subcontractor wins the case. The wording of the clause states:

> 'the party serving the Notice to Adjudicate shall bear all the costs and expenses incurred by both parties in relation to the adjudication, including but not limited to, all legal and expert fees'.

It was argued on behalf of the subcontractor that the clause by its very nature inhibited parties from pursuing their remedies under the Act and was therefore void. This being the case the adjudicator had exceeded his jurisdiction by applying the clause.

The court held that the clause was valid. The mere fact that, in this particular case, the subcontractor was disgruntled did not entitle it to successfully argue that the clause was unfair and should not apply.

SUMMARY

Despite the unfair nature of such a clause, the courts will take the view that, if parties are prepared to enter into contracts which include clauses of this kind, they will be enforced.

TABLE OF CASES

SUBJECT INDEX